Alarm Management

Seven Effective Methods for Optimum Performance

Alarm Management

Seven Effective Methods for Optimum Performance

by
Bill R. Hollifield
PAS Principal Alarm Management Consultant
&
Eddie Habibi
PAS Founder and CEO

Foreword by
Ian Nimmo
President and a Founder of User Centered Design Services

67 Alexander Drive
P.O. Box 12277
Research Triangle Park, NC 27709

Printed in the United States of America.
10 9 8 7 6 5 4 3 2

ISBN-13: 978-1-934394-00-7
ISBN-10: 1-934394-00-9

Notice

The information presented in this publication is for the general education of the reader. Because neither the author nor the publisher has any control over the use of the information by the reader, both the author and the publisher disclaim any and all liability of any kind arising out of such use. The reader is expected to exercise sound professional judgment in using any of the information presented in a particular application. Additionally, neither the author nor the publisher have investigated or considered the affect of any patents on the ability of the reader to use any of the information in a particular application. The reader is responsible for reviewing any possible patents that may effect any particular use of the information presented.

Any references to commercial products in the work are cited as examples only. Neither the author nor the publisher endorses any referenced commercial product. Any trademarks or tradenames referenced belong to the respective owner of the mark or name. Neither the author nor the publisher makes any representation regarding the availability of any referenced commercial product at any time. The manufacturer's instructions on use of any commercial product must be followed at all times, even if in conflict with the information in this publication.

Library of Congress Cataloging-in-Publication Data

Hollifield, Bill R.
Alarm management : seven effective methods for optimum performance / by Bill R. Hollifield, Eddie Habibi ; foreword by Ian Nimmo.
p. cm.
Includes bibliographical references.
ISBN 978-1-934394-00-7
1. Process control—Automation—Handbooks, manuals, etc. 2. Detectors—Handbooks, manuals, etc. I. Habibi, Eddie. II. Title.
TS156.8.H66 2007
681'.25—dc22

2007004426

For Pam, my beautiful, patient, and adorable trophy wife of 29 years, whom many people have asked "Why did you marry outside of your species?"

—Bill

To the scientists, engineers and industrial plant operators around the world whose creativity and perseverance have resulted in product innovations and manufacturing processes that have improved the quality of life for mankind more in the past 150 years than any period since the beginning of recorded history.

To my team at PAS, for their ingenuity, hard work and fun-loving attitude.

—Eddie

In Memoriam

Chris Uehlinger
1956–2005

In 2005, PAS lost President and Chief Operating Officer, Chris Uehlinger. With his death, PAS has lost a great friend and business partner, and the process controls industry worldwide has lost an admired industry leader and visionary.

Chris' commitment to excellence helped shape the vision of improved process safety and reliability through advanced automation technologies.

Chris joined PAS in the spring of 2000 after a long and successful career as a vice president with Honeywell Process Solutions. At PAS, he continued to demonstrate the qualities that made him an exceptional leader: thoughtful and strategic thinking, personable leadership, and an extraordinary ability to grasp and communicate complex technological concepts. Chris' career began at Celanese after he earned his engineering degree from Texas A&I University (now known as Texas A&M University) in Kingsville, Texas. Chris will be greatly missed and fondly remembered.

—Eddie Habibi, PAS Founder and CEO

Table of Contents

List of Figures

List of Tables

Foreword

In the rush to take advantage of computer-based automation, many companies in the processing industries worldwide have overlooked one of the most important individuals in their business value chain—the plant operator. The plant operator is *the forgotten knowledge worker.* He is on the frontline of real-time operations, making decisions that directly impact plant safety, reliability, profitability, and ultimately shareholder value.

Over the past three decades, distributed control systems and layered applications have enabled most processing plants to expand the scope of responsibility of the plant operator while significantly increasing the sophistication and the complexity associated with operating a plant. By any measure, the control room operator today is a *knowledge worker,* not too dissimilar from an airline pilot, a doctor or a plant manager. What makes these individuals knowledge workers, according to Peter F. Drucker, the legendary business management guru, is that they all have large amounts of information as input into their work processes. Operators—like other knowledge workers—analyze information, diagnose situations, predict outcomes, and take action to deliver value. And in the case of the airline pilot and the plant operator, their real-time response has a much more urgent meaning to it than the stereotypical knowledge worker.

Today there are no unified standards of competency applied to plant operators. Yet manufacturers continue to increase the scope of responsibility of the operator by consolidating their positions in the name of efficiency, or profitability. There is certainly no argument here against reducing headcount and improving shareholder value. The case for debate here is whether we are doing enough to retool the residual operator(s) to ensure safe and reliable operation of the plant. We maintain that today only a few operating companies take adequate measures to retool their operators to compensate for the reductions in force. And not surprisingly, retooling the operator has been proven to increase his ability to mitigate accidents, from minor to catastrophic.

The anatomy of an industrial accident encompasses many facets and is hardly ever the result of a single malfunction. While there may be a single root cause that triggers an accident, such root causes are rarely the only factor in propagating the accident. As discovered recently at a refinery in the US Gulf Coast, accidents generally are the result of multiple failures within a relatively short period of time. Displays fail to show proper indications.

Sensors fail to report correct measurements. Important alarms fail to enunciate or are lost in the flood of other alarms. Operators make the wrong move. Mechanical relief systems fail. And so on and so on.

The severity of consequences of accidents is also directly impacted by the operator's ability to recognize the initial abnormality, his ability to assess and diagnose the situation, and finally, the timeliness and the accuracy of the actions he takes. Hence, situational awareness and rigorous vigilance in the control room by competent operators are important factors in mitigating abnormal situations.

In other cases, such as the Texaco Pembroke Refinery in the UK, combinations of factors played a major role. In this major accident, bad weather initiated the power outage and subsequent shutdown of the FCC unit. Following that, there were failures of engineering and operations teams to follow proper MOC processes for mechanical changes in the piping of the unit, and poor judgment by management in pressuring the operations team to startup the plant. In the DCS, there was the lack of an overview display, inaccurate display of a valve position, and an overwhelming barrage of alarms. These and several other factors combined and led to an explosion that caught the attention of operating companies as well as health and safety authorities around the world. The one "lucky" factor that precluded the Pembroke refinery accident from becoming a major industrial disaster was the fact that it occurred on a Sunday and most plant personnel were either not at work or inside the control room.

Many of the accidents during the past twenty years may have been prevented, or their consequences minimized, had closer attention been given to the role of the operator and the factors that surround him. These factors are collectively known as *situational awareness, operations effectiveness solutions, or critical condition management.* These important factors for ensuring plant safety, reliability and profitability include but are not limited to the following areas:

- operator interface
- alarm management system
- regulatory control loops
- control room human factors including lighting, comfort, communication, unit console adjacency, etc.
- simulation-based operator training

- automated procedures
- automated transition management
- early fault detection and operator advisory
- online knowledge support systems

We estimate that most of the world's processing plants lack several of these operator effectiveness solutions.

Visions of process automation grandeur have been around for quite some time. They include "lights-out" factory floor and "operator-less" remote operations concepts. Such visions are not entirely unrealistic. In fact, as of the publication of this book, we believe adequate computing power and advanced technologies exist to achieve a large portion of such visions. One of the biggest challenges to achieving a totally automated manufacturing environment is the disparate nature of the available technologies.

Over the years, different automation suppliers have developed various components of the puzzle creating silos of automation, most of which share no common components. Many operating companies have also treated the automation opportunity/challenge as standalone endeavors desiring, rightly so, for the best-in-class solution in every area. Typically, one group in the company focuses on upgrading the control system, another on advanced controls and optimization, and others on instrumentation, safety systems, and information technology. No one group seems to be concerned about how the integration of these systems will impact the operator.

To make matters worse, some operating companies went to engineering and construction contractors or DCS suppliers for their control computer upgrade projects, who in turn treated the system configuration, the operator user interface, and the alarm management design as commodity work. In many cases, the same engineer in charge of field instrumentation (who had no training in human factors engineering) ended up being responsible for designing the applications that directly affect operator effectiveness. DCS suppliers provided plenty of alarms, in some cases up to sixteen per sensor, and engineers felt compelled to use them all without any rationalization. What the industry ended up with, in many cases, was an alarm system that became a hindrance to the operator instead of being a useful tool, especially during abnormal conditions.

Now, consider the case of the safety instrumented system or advanced process control. Companies in general recognized that these are special fields requiring professionals with specialization and experience in these areas. Unfortunately, the same consideration was not applied to the very factors that directly impact the most important individual in the value chain of every processing plant—the operator.

While the correct approach to optimizing safety, reliability and profitability of a plant through automation requires an overall consideration of all the key factors of operator effectiveness, it is generally a significant effort to direct the attention and the commitment of an organization to take on such initiatives. Most companies still deal with these issues as standalone problems.

Meanwhile, the reality of day-to-day plant operation dictates that immediate steps be taken to improve the operator's vigilance and ability to manage abnormal situations, however incremental such efforts may be. It is with that reality in mind that this book takes on one of the several steps toward improving operations effectiveness of a processing plant.

In recent years, alarm management improvement has become recognized as an immediate "low hanging fruit" in improving operator effectiveness. This awareness is mostly due to the efforts of industry organizations such as the ASM Consortium® and EEMUA as well as the efforts of advanced automation solution providers like PAS, UCDS, and others.

The most pressing question for some companies today, however, is how to justify alarm management improvement projects. This is in spite of the fact that many industry leaders have already identified improving the alarm system as a corporate mandate requiring no further cost justification at the individual plant level.

This book is the most comprehensive compilation of the alarm management improvement body of knowledge. The methodology for improving an industrial alarm system captured in this book has been defined, refined and field proven over the past decade. The methodology benefits from hundreds of projects that have yielded quantifiable breakthrough results. The reader will be able to use the prescribed steps outlined in this book to improve any alarm system regardless of the brand or the condition of the system.

The approach outlined in this book is influenced by the principles of Six Sigma quality work process and project methodology. Special effort was made to not encumber the reader with nonessential aspects of Six Sigma. In fact and in practice, the process defined in this book benefits from Six

Sigma without requiring much of the dreaded overhead typically associated with Six Sigma projects.

Back to the *forgotten knowledge worker,* we believe that by retooling the operator with proper situational awareness solutions such as an effective alarm management system, user-centered designed operator interfaces, effective training and more, industrial companies can begin closing the gap between their *knowledge worker's* effectiveness and that of the airline industry's and others.'

To the practitioner of alarm management this book will provide the practical examples, project management skills and engineering discipline to make future alarm projects successful and not wasted effort far beyond today's initial alarm database configuration clean-up and will enable operators to be all that they can be.

> *"Anything that is wasted effort represents wasted time. The best management of our time thus becomes linked inseparably with the best utilization of our efforts."*
>
> *—Ted W. Engstrom, Topic: Existence*

—Ian Nimmo
President and a Founder of User Centered Design Services

Ian Nimmo

Ian Nimmo is President and a Founder of User Centered Design Services, an ASM Consortium® affiliate member and an ASM service provider. He served 10 years as a Senior Engineering Fellow and a founder and Program Director for the ASM Consortium®. He also has 25 years prior experience as an engineer for Imperial Chemical Industries (ICI) in the UK. Ian has written over 100 papers and contributed to several books on the subject of design for operator safety and abnormal situation management. Ian holds a degree in electrical and electronic engineering from Teesside University in the UK.

Acknowledgments

Many people have significantly contributed to the Alarm Management body of knowledge and deserve recognition and thanks for those contributions. Here are a few of these people.

The PAS Team

First and foremost, the authors greatly appreciate the significant effort, quality of work, and patience Jennifer Hicks demonstrated in supporting and driving the team to deliver on a very tight schedule.

The following people are current or former PAS experts in DCSs, alarm rationalization, state-based alarming, and programming who have spent many years on the front lines of developing and delivering these solutions to customers.

Dennis Ahrens	Jerry L. Brauner	Nick Cappi
J. Patrick Connor	Robert Croy	David A. Day
Dan Dearwater	John Fairbanks	Jerry K. Gooch
Bijan Habibi	Tom Hosea	Robert S. Ketcham
Samir S. Kulkarni	Rajeev Limaye	Michael Morano
Christina Ng	Jacques Smuts	John A. Soto
John Stritch	Henry To	James B. Ward
Erik Zanker		

John M. Wright—a dedicated, world-traveling early pioneer in alarm philosophy development and rationalization. An expert in DCS configuration and programming. And an all-around great guy that we have known for many years.

Significant Contributors to the Alarm Management Body of Knowledge

The following gentlemen have provided many knowledgeable publications and articles, and have significantly helped shape the field of Alarm Management.

Donald C. Campbell-Brown, BP Exploration Operating Co. Ltd, Sunbury, UK

Ian Nimmo, User Centered Design Services

Mathew Bransby (Deceased)

Company Champions, Consulting Practitioners and Specialists in Alarm Management

Peter Andow	Jim Cawood	Mike Clark
Kenneth F. Emigholz	Ralph Harris	Heinz Janiec
Les Jensen	Yahya Nazer	Douglas Rothenberg
Nicholas P. Sands	Bill Sheldon	David Strobhar
Lee Swindler		

Organizations

EEMUA: The Engineering Equipment and Materials User Association

The Abnormal Situation Management (ASM®) Consortium

About the Authors

Bill R. Hollifield, PAS Principal Alarm Management Consultant

At PAS, Bill is the Principal Consultant responsible for the Alarm Management work processes and products, intellectual property, and software product directions. He is a voting member of the ISA-SP18 Alarm Management committee.

Bill has international, multi-company experience in all aspects of Alarm Management. Besides that, Bill has many (many!) years of chemical industry experience with focus in project management, chemical production, and control systems.

Bill holds a Bachelor's Degree in Mechanical Engineering from Louisiana Tech University and an MBA from the University of Houston. He's a pilot, and builds furniture (and the occasional log home in the Ozarks) as a hobby.

Eddie Habibi, PAS Founder and CEO

Eddie Habibi is the Founder and CEO of Houston-based PAS. Prior to establishing PAS, Eddie held various positions at Schlumberger and Honeywell International. Eddie holds an engineering degree from the University of Houston and an MBA from the University of St. Thomas.

Eddie has led the growth of PAS since its initial founding in 1993 as a professional consultancy to being a global provider today of advanced automation solutions to the processing industries worldwide with operations in the United States, Europe, the Middle East, Africa and Asia.

Eddie is a recognized industry thought leader in the areas of operations effectiveness, automation intellectual property asset management, data mining and knowledge management. Under his leadership PAS continues to grow profitably as an agile organization known for its strong culture of customer focus and significant innovation.

INTRODUCTION

"There are more things to alarm us than to harm us, and we suffer more often in apprehension than reality."

—Lucius Annaeus Seneca

Why We Wrote This Book

The purpose of this book is to capture in one volume the current body of best practices knowledge for improving and optimizing the performance of a modern alarm management system.

This is an intentionally different kind of book than you usually find in engineering circles. It is certainly a comprehensive and complete overview of Alarm Management Optimization strategies for distributed control systems. However, the main focus is on practical advice and techniques to significantly improve the performance of existing alarm systems. The principles herein can also enable new systems to be initially configured correctly, and not require expensive re-engineering after problems later become apparent.

The book is based upon more than a decade of extensive experience in the Alarm Management Optimization (AMO) field including experience in almost every industry. All manner of practices make up the background information used in producing this book—the good and the bad, the best and the worst. The basis includes a working knowledge of the guidelines, articles, reference works and other materials on the subject, along with knowledge and experience obtained from hundreds of alarm management improvement projects. Breakthrough results have been achieved by following the principles contained in this book.

We believe it is time to put, in one place, the current body of knowledge on the subject, tailored specifically to the solution of the problem. The original reference works on the subject are a good starting point, but we hear repeatedly that they are rather topical and none address the entire alarm management problem holistically. Also, since their creation several years ago, many terabytes of actual system alarm data have been gathered and analyzed, and much of the advice and methodologies mentioned earlier need updating and refinement to reflect current real-world conditions.

In this book, you will find actual examples of good practices and poor practices. The various problems of alarm systems are covered with precise guidance on how they come about and how to effectively correct them. We know that most operating companies are limited in time, money, and resources. We will not advocate academic, theoretical, or impractical approaches to the problems. Instead, in this book you will find fact-based, field proven, straightforward and practical solutions.

Is This Book for You?

This book specifically targets alarm management related to modern Distributed Control Systems (DCSs). These flexible and capable systems are used throughout various industries, including oil and gas, refining, chemical, petrochemical, pulp and paper, pharmaceuticals, power generation, minerals processing, discrete manufacturing, and others. The most common scenario this book addresses is a processing facility—continuous or batch—with one or more operators using a modern DCS. If you have such a facility, you will find this book valuable.

This book also serves those involved in designing and modifying such facilities. Proper alarm management practices are most effectively and inexpensively accomplished in the original specification, design, and configuration of a control system. Many companies now require the application of these principles in the design phase; this book will tell you how to do it right the first time.

CHAPTER 1

Alarm Management Best Practices: Highly Condensed

"My definition of an expert in any field is a person who knows enough about what's really going on to be scared."

—P. J. Plauger

1.1 The Alarm Problem

A poorly functioning alarm system is often noted as a contributing factor to the seriousness of upsets, incidents, and major accidents. Significant alarm system improvement is needed in most industries that utilize computer-based distributed control systems; it is a massively common and serious problem. Most companies have become aware that they need to thoroughly investigate and understand their alarm system performance. Alarm management is a fast-growing, high profile topic in the process industries. It is the subject of constant articles in the trade journals and at various technical society meetings and symposia.

Having decided to investigate this area, how do you proceed? Your time and resources are always limited. The subject is complex. Alarm system improvement involves an interlinked combination of technology and work processes.

1.2 People Who Can Help

You should seek help from the best experts in the field. You will likely want information, advice, products, and services from:

- People who are acknowledged experts in the alarm management field with in-depth understanding of the historical and current problem, the science and literature, the studies, and the range of solutions.
- People with in-depth knowledge of process control, distributed control systems, human-machine interfaces, process networks, and critical condition management.
- People with experience in every stage of a successful alarm system improvement project, along with many examples of successful projects.
- People who know work processes, based on successful experience in different industry segments. You want to know what the industry is doing, what are the best and most efficient practices, and frankly, what are the worst practices.

1.3 Seven Steps to a Highly Effective Alarm System

Here is a very brief outline of a best practices approach in a typical alarm management project. These straightforward steps can be easily implemented in any work process framework, such as Six Sigma. The first three steps are universally needed for the improvement of an alarm system. They are often done simultaneously at the start of a project.

"Always-needed" steps:

Step 1: Develop, Adopt, and Maintain an Alarm Philosophy

Step 2: Collect Data and Benchmark Your Systems

Step 3: Perform "Bad Actor" Alarm Resolution

> *Steps to implement based on alarm system performance after the first 3 steps:*
>
> Step 4: Perform Alarm Documentation and Rationalization (D&R)
>
> Step 5: Implement Alarm Audit and Enforcement Technology
>
> Step 6: Implement Real Time Alarm Management
>
> Step 7: Control and Maintain Your Improved System

Step 1: Develop, Adopt, and Maintain an Alarm Philosophy

An Alarm Philosophy is a comprehensive guideline for the development, implementation, and modification of alarms. It provides an optimum basis for alarm selection, priority setting, configuration, response, handling methods, system monitoring, and many other topics. In this book you will learn exactly how to develop an Alarm Philosophy, with examples. An Alarm Philosophy will be an immediately useful document that will cover the entire range of alarm topics. It will reflect a full understanding of the alarm problem and the proper practices to follow.

Step 2: Collect Data and Benchmark Your Systems

Analysis is fundamental to improvement. You must analyze your alarm system to improve it. You should look for alarm analysis software with full graphical and tabular output, easy access to the full DCS journal entries, automatic report generation, web-based report viewing, and so forth. You want a comprehensive and complete set of alarm analyses to enable you to pinpoint your exact problems and apply the most efficient solutions.

Since operator changes (controller setpoints, modes, and outputs, for example) are recorded by most DCSs in a similar fashion to alarm events, you will want software that includes the analysis of such events. The results can be amazingly useful, and point out areas where control schemes are not working as designed or where operating procedures or operator training need improvement. While this book is focused on Alarm Management, we include a section on the benefit of these operator change analyses.

There can be no improvement without an understanding of your starting point. A comprehensive Baseline Report sets your benchmark—and will enable you to target your resources to get the most improvement possible for the minimum cost and effort. The start of an improvement effort requires an examination of your actual data.

Step 3: Perform "Bad Actor" Alarm Resolution

Based on the analysis of hundreds of systems, there are *always* several varieties of nuisance or "bad actor" alarms. This book contains an efficient and effective process for analyzing these and providing exact recommendations for configuration changes to improve their performance. The average improvement is over a 50% reduction in overall alarm events for a relatively minimal effort. While on some systems this result may not meet an overall improvement goal, it is a great first step, providing much-needed immediate relief.

The previous three steps are universally needed for the improvement of an alarm system. The following steps may or may not be needed depending on the performance characteristics of your system. And since these can be expensive and time-consuming steps, you want to be sure that they are needed before you commit to them.

Step 4: Perform Alarm Documentation and Rationalization (D&R)

Many systems need a total rework—a review of the configuration and purpose of every alarm. We call this Alarm Documentation and Rationalization (D&R), also commonly called Alarm Objective Analysis, among other terms. You will want to use a software-assisted methodology to make D&R fast and efficient. Besides just having software, there is an "art" to performing a D&R in an efficient manner. The knowledge herein is based upon participation in the rationalization of hundreds of thousands of points. This experience provides detailed knowledge of the common problems and the best solutions, which are provided here in this book. One result of a D&R effort is the creation of a Master Alarm Database, which contains the post-rationalized alarm configuration with changed setpoints, priorities, and so forth. A Master Alarm Database has several uses.

Step 5: Implement Alarm Audit and Enforcement Technology

Once your alarm system is improved, it is essential to ensure that the configuration does not change over time unless the changes are specifically authorized. DCS systems are notoriously easy to change, which is why software mechanisms that frequently audit (and enforce) the current configuration versus the Master Alarm Database are needed. Paper-based Management of Change solutions for DCS configuration (alarm or otherwise) have a wide and consistent history of failure.

Step 6: Implement Real-Time Alarm Management

Based on the performance level you need your alarm system to achieve, you may want to implement safe and flexible state-based alarming, alarm flood suppression, and alarm shelving capabilities on your system.

- **State-based Alarming:** Algorithms detect when the plant changes operating state (such as startup, shutdown, rates, different feed-stocks, etc.) and dynamically alter the alarm settings to conform to the proper settings for each state. State-Based settings for inadvertent shutdown of a piece of equipment have proven to be very effective in managing most alarm flood situations.
- **Alarm Shelving:** A safe, secure way to temporarily disable a nuisance alarm until the underlying problem can be corrected. Computerized lists of shelved alarms, with reminders and auto-re-enabling, are necessary. It must be impossible to temporarily suppress an alarm and then "forget about it"—a very common and very dangerous occurrence throughout industry.

Step 7: Control and Maintain Your Improved System

Processes and sensors change over time, and alarm behavior will change with them. Alarms that work correctly now may become nuisances or malfunction in the future. An ongoing program of system analysis, and the correction of problems as they occur, is needed for an effective alarm system.

1.4 Summary

If you know or suspect that you have an alarm problem, read this book and begin doing the things it recommends.

CHAPTER 2

The History and Nature of the Alarm Problem

"We live in a moment of history where change is so speeded up that we begin to see the present only when it is already disappearing."

—R. D. Laing

2.1 How Did We Get In This Mess?

Like many problems, this one began with the best of intentions. In the "good old days," a control room had a wall full of individual process indicators, lights, switches, and moving-pen charts. These items took up a lot of room, which was always in short supply. One particular item of interest to us was the alarm system—a rectangular array of labeled windows that individually lit up (usually flashing) based on their particular connection to the process. This "lightbox" also incorporated a horn that would sound when an alarm activated, and an Acknowledge button to silence the horn and change the flashing light to a steady light. This Acknowledge button was also often equipped, by the end user, with a wedge of paper or coin to hold it in and keep the infernal noise from happening in the first place. Certainly in place on night shift, this enhancement might get removed during the day.

The "control wall" concept had many positive things going for it. Considerable thought went into placement and grouping. Trends were always visible as long as the paper and ink were replaced. The overall health of the process could be ascertained almost at a glance. The alarm display would often produce repeatable patterns depending on the type of upset (see Figure 2–1).

The disadvantages to these systems were many. Inter-controller connectivity was almost non-existent. The implementation of complicated control schemes was difficult. Introduction of new controls involved either an expensive relocation of adjacent elements, or the sacrifice of their logical placement. Communication of the control system to "anything else" was generally impractical.

Regarding alarms, the addition of a new alarm was expensive. Their total number was limited by space availability and cost. Therefore, each one was individually evaluated and justified.

This was the situation prior to the digital revolution and the introduction of the Distributed Control System (DCS). The business advantages of a DCS are huge! It is easy to reconfigure control strategies. It is easy to programmatically alter the behavior of the system. And almost everything in the system is changeable without much trouble.

So, over a couple of decades, DCSs have predominated, and most older-style control systems have been converted. The advantages of a DCS have far outweighed some well known deficiencies that are outside the scope of this book.

The situation for alarms is far different in a DCS than in an older system. Since alarms are basically displayed on a computerized scrolling list or on graphics, there is unlimited space for them—at least from the point of view of the DCS manufacturer, if not the operator. And since every point in the DCS is essentially a software construct, alarms are "free." Most points in the DCS have several possible alarms just waiting for the engineer to configure and activate by touching a few keys. No justifying, no wiring, no tubing, no plastic engraving—just click, click, click and you have a new alarm.

Figure 2–1 A "control wall" with alarm lightboxes at the top.

And create them we all did! With no consistent guidelines to follow, massive over-configuring of DCS alarms is common. After all, if the manufacturer supplied the functionality of a High, High-High, and even HHH alarm, well then, they must be there for a good reason, so let's use them all!

With no guidelines or cost for creation of alarms, poor practices arise—such as all alarms enabled by default, set up by inconsistent rules of thumb, or set by an individual's preference. Consistency was low; similar process systems implemented by different teams would have significantly different alarm configurations and behavior.

The result? The well known "Alarm Problem" (see Figure 2–2) wherein, even in steady-state process operation, the alarm system is activating almost continually, creating far more alarm events than can be possibly individually understood and acted upon by the operator. During an upset, there is an order of magnitude increase in the number and speed of alarm occurrences, rendering the alarm system not only useless, but *creating an active hindrance* to the operator's ability to deal with the situation. Time and time again, investigative reports after major industrial accidents have

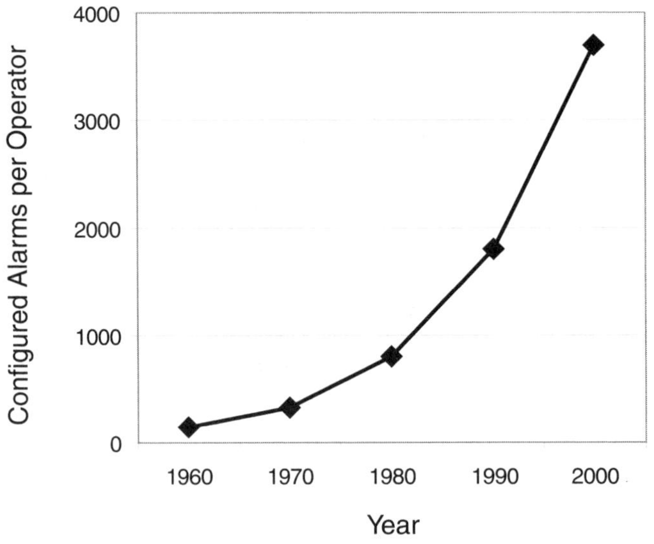

Figure 2–2 The number of configured alarms per operator has increased exponentially.

shown that overloaded, bypassed, or ignored alarm systems have played a significant role in making the situation worse.

And the major accidents are just the tip of the iceberg. It is also well known that ordinary process upsets can be made either worse or made to last longer if they are associated with an ineffective alarm system. There are very large costs associated with this phenomenon.

The situation is made even worse by the ease of modifying alarms in a DCS. Not only can engineers change the alarm configuration, but so can operators, maintenance technicians, and college interns. Or worse, even managers! Security of such change is generally via key or password, and everyone that is knowledgeable assumes that every operator has such access. Since the 1990s, manufacturing sites have had rigorous Management of Change (MOC) policies to address almost any change you can think of in a process facility—except for the alarms! For many years, the configuration, alteration, and bypassing of alarms in a DCS has often been ineffectively covered by MOC policies and practices.

The result is alarm systems that are overloaded and ineffective. Such systems are common throughout industry.

2.2 Where Are We Now?

The alarm problem began to be identified and written about in the early 1990s. Some major industrial accidents occurred, and the alarm systems were written up as contributory factors in the reports.

An example from the Health and Safety Executive (HSE) report on a 1994 major refinery accident:

- There were too many alarms and they were poorly prioritized.
- The control room displays did not help the operators to understand what was happening.
- In the last 11 minutes before the explosion the two operators had to recognize, acknowledge, and act on 275 alarms.

A variety of articles were written on the topic of alarm management, and several companies began to offer various products and services to address the issue. Alarm rationalization began with spreadsheets and overhead projectors, and evolved into dedicated software specifically designed for the task. Dynamic alarm management software was introduced.

The Abnormal Situation Management (ASM®) Consortium was formed in 1994 and began studying aspects of the problem, and acted to greatly increase awareness of it. In 1999, the Engineering Equipment and Materials Users Association (EEMUA) produced a seminal reference document (their Publication 191) on the topic.

2.3 OSHA's Position on Alarm Management

An overlooked fact in the industry is that OSHA (the United States Occupational Safety and Health Administration) specifically outlines certain requirements for alarm systems in CFR 29, regulation 1910.119, *Process Safety Management of Highly Hazardous Chemicals*. These requirements are related to the accurate documentation of critical alarms, and operator training for understanding and responding to alarms.

The regulation requires that alarms be documented and the information made available to operators with clear indications of:

- operating limits
- consequences of deviation
- steps required to correct or avoid deviation

Alarm documentation for operator training is a law in the United States.

For many companies, Alarm Management has become an accepted element of good operations management. Industrial insurance companies are beginning to require their clients to implement alarm managment practices, and companies are assigning resources and phasing in programs to do so. The issue has changed from "study and evaluation" into action.

CHAPTER 3

How Do You Justify Alarm Management?

"Anyone who says businessmen deal in facts, not fiction, has never read old five-year projections."

—Malcolm Forbes

Proper alarm management will result in improved safety, reliability, and overall profitability. It is the right thing to do, especially in light of the recent rush to reduce operational costs through reductions in head count. Those that wait to be told by regulatory authorities to address the alarm management problem will miss out on a competitive advantage in the meantime.

It is difficult to rigorously cost-justify an alarm improvement project. The Abnormal Situation Management (ASM) Consortium estimates that in the US alone over 20 billion dollars per year are lost due to abnormal situations. Inadequate alarm systems contribute to a considerable portion of that loss, and it can be argued that many of these abnormal situations could be mitigated with robust alarm management systems. However, trying to go from that large number to a hard-dollar justification for a specific facility is very difficult. Certain operating companies have internally identified alarm management improvement as a business imperative and operational best practice. One such company has correlated their plant performance, including production rate, to alarm system performance and operator loading. This relationship is accepted by that company as significant, and they take the improvement of their alarm systems seriously.

In trying to help appreciate the challenge of identifying cost justification arguments for alarm management projects, we will make a contrast and an analogy here.

First, the contrast. Advanced process control (APC) directly impacts key indicators (KPIs) of the economic performance of a plant. APC improves production and quality, as well as reducing off-spec materials. These key performance indicators are easily measurable. The beneficial results of APC are clearly shown by comparing those KPIs before and after APC implementation.

Measuring the direct contribution of alarm management improvement is more difficult, as there are so many other factors that are tightly related to the performance of a processing plant. These factors include equipment maintenance, control system performance, the operator user interface, operator competency, and others.

Sound alarm management is more about mitigating—preventing and minimizing the impact of—abnormal situations, and less about direct production throughput improvement. When everything is going well, a good alarm system does not move the valves to make more or better products. One will, however, significantly influence the operator's ability to reduce the consequences of an emerging abnormal situation.

Therefore, the alarm system should be viewed in the same light as the Safety Instrumented System (SIS), in that they are both important and essential layers of protection for preventing unwanted incidents (see Figure 3–1).

Proper mitigation of even small incidents will save money. For every shutdown that a plant experiences in the course of a year, there may be four to five times as many minor incidents with associated rate reductions. The impact of these small incidents and their cumulative costs are often overlooked, when calculating the impact of abnormal situations on plant profitability.

In Figure 3–2, the graphical point is made that even a few minor upsets can totally wipe out the additional profit that is made by investing in

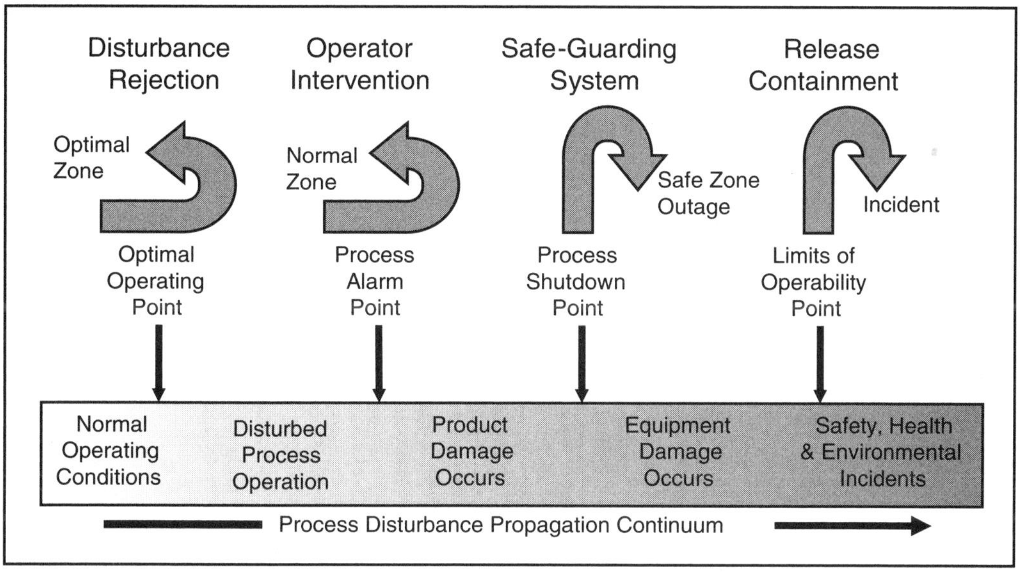

Figure 3–1 Process disturbances and layers of protection.

APC. Thus the avoidance of such upsets, or the lessening of their intensity or duration, has a payback. Poorly performing alarm systems do not cause accidents, upsets, and rate reductions. However, it is well-documented that they can and do make such incidents worse, or make them last longer than they would otherwise, and that does result in the loss of considerable money.

Safety Instrumented Systems (SISs) have now become an accepted and mandated requirement for processing plants. Such systems often consist of expensive, fault-tolerant, high-performance programmable logic controllers. These systems use redundant inputs and outputs and incorporate fail-safe technology. They are monitored and maintained to result in extremely high levels of reliability and risk mitigation. The analogy for alarm management justification has to do with such safety systems.

So, imagine that you hire an expert to evaluate such a system, say, an expensive "Tri-Safety 5000". The expert proves conclusively that your Tri-Safety 5000 is in covert failure mode for 15% of the time. During those periods, it is not affording you any protection at all. What would your reaction be?

At most places, the reaction would be as follows: At the very least, the CEO of Tri-Safety would be immediately informed, lawsuits threatened, and their expert team of engineers would be camped out in your facility

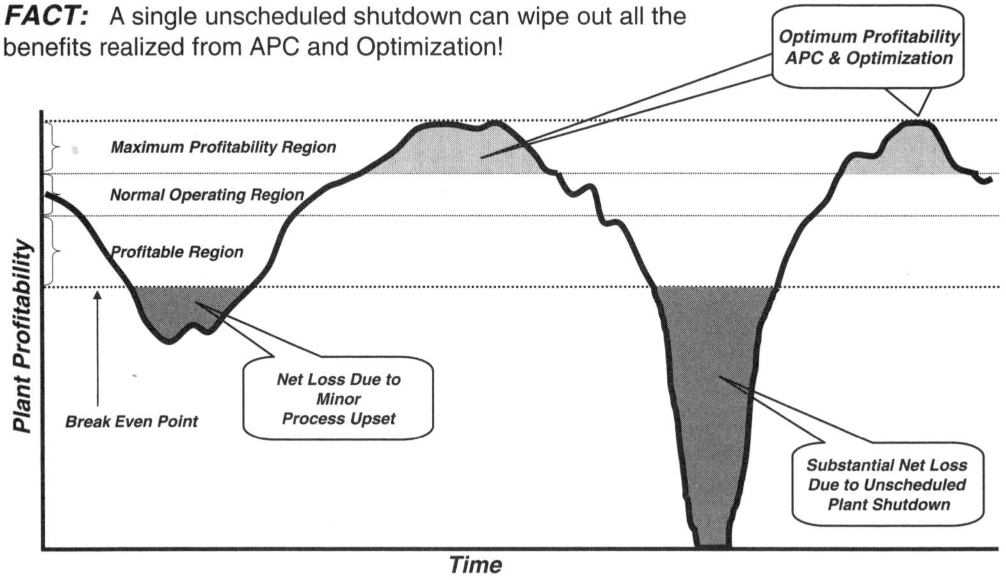

Figure 3–2 Process upsets and profitability.

and not leave until they made the thing work all of the time! Which is why you bought it in the first place. There would be no excuses. In some processes, you would probably perform a total shutdown until the Tri-Safety 5000 was fixed.

If it is agreed that the alarm system, similar to the SIS system, is a layer of protection, then its performance must be taken as seriously as that of the SIS. The analogy to the alarm system is direct. It is straightforward to calculate and prove that a poorly-performing alarm system is essentially not available for much of the time. Alarm analysis can pinpoint those periods and measure their magnitude and duration. There are many alarm systems that would be of more use to the operator if they were totally turned off rather than continuing to be the nuisance and distraction that they currently are. The alarm system is there to mitigate risk and ensure that abnormal situations are made known to the operator so that they can be corrected. If it is not working, you have lost that protective capacity.

You can easily prove, on many systems, that the operator gets so many alarms that hundreds or thousands per day are being ignored—because the operator has *no choice* but to ignore them. It is a physical impossibility to properly analyze and respond to the alarm rates that are commonly seen in industry. Such alarm systems are not working. As a facility owner or manager, are you comfortable with that? What assurance do you have that the operators are ignoring "the right ones"? If you have an incident, what is your exposure if your alarm system is not performing its proper function for much of the time?

And here is a further reason "ripped right from today's headlines." Besides the fines and financial penalties levied against a company after an incident, now and in the future you may well see *criminal charges* levied against plant management. The failure to manage your alarm system could well end up in charges of criminal negligence. This is a new development. Imagine taking all of your personal retirement savings and spending them on a defense attorney, if your company won't cover those costs—as well it may not.

We believe that, while there may be a way to directly relate the performance of an alarm system to the bottom line, it is a very tedious effort and at the end, the results may be contentious. There are simply too many variables that cloud the metrics. For many operating companies the question of alarm management cost justification has been long resolved and it has become a standard best practice. They have made it a corporate mandate, and require every plant to implement it as they would require the installation and proper maintenance of a safety system.

We further believe that sooner rather than later, in most industrial countries, government agencies in charge of safety and environmental matters will address this issue. They will institute alarm management best practice standards, as practiced by industry leaders, as a regulatory requirement. This seems to be the trend.

In today's environment, proper configuration and management of your alarm system is not an option. It is a requirement. It is part of the cost of doing business.

CHAPTER 4

What Should Be an Alarm?

This is the most important chapter in this book.

The decision to configure an alarm must meet several specific criteria. The most important of these criteria is:

> ***Does the event require operator action?***
>
> ***Working Definition: A process alarm is a mechanism for informing an operator of an abnormal process condition for which an operator action is required. The operator is alerted in order to prevent or mitigate process upsets and disturbances.***

Alarms everywhere are configured without meeting this criterion, which is one of the main reasons that the alarm problem exists. Because DCS manufacturers have made the alarm system so easy to use, it is used for many inappropriate things.

Without an agreement as to this working definition, which is the basic principle of proper alarm configuration, it is unlikely that any improvement in an alarm system will be achieved. But if this principle is followed, huge improvement in a system will be made, even if none of the other principles in this book are followed—it is that powerful.

The alarm system must be *reserved* for events that require operator action. Only such events shall be configured as alarms. Events that are useful only "for operator information" or similar reasons, not involving action, can and should be presented in a variety of ways other than the use of the alarm system, as we will discuss later.

So, what constitutes an "operator action" for these purposes? The following are appropriate:

- Making process changes by manipulation of the control system
- Directing others to make changes in the control or process system (manually start pumps, operate valves, check items for function, take samples, etc.)
- Beginning troubleshooting and/or analysis of a situation
- Increasing the monitoring of a section of the process
- Contacting other people or functional groups regarding a situation
- Logging conditions for the necessary purposes of later examination, maintenance, or repair
- Changing operating mode because of a difference in system capability; for example, running things manually because an automatic subsystem has failed, or running further from optimum because the Advanced Process Control system has failed

What are *not* "operator actions" for the purpose of validating an alarm? Here are some examples:

- Writing something down in a logbook (other than preparing a maintenance work order!)
- Thinking "OK, that's nice to know."
- Thinking "OK, the next shift can deal with that tomorrow."
- Thinking "OK, the system is working normally."

Alarms must exist solely as a tool for the benefit of the operator. They are not to be configured as a tool or for the benefit of the control engineer or other staff.

So, the most important criterion for an alarm is the requirement for operator action in response. There are a total of three major criteria:

1. Does the event require operator action?

   Events that do not require operator action *shall not be allowed to produce alarms*. It is common historical practice in industry to ignore this principle.

2. Is this alarm the best indicator of the situation's root cause?

   Alarms should be placed, configured, and handled so that a single process event does not produce multiple alarms all signifying the same thing. That is, alarms should be configured on the best indicator of the root cause of a situation.

3. Is this alarm truly resulting from an abnormal situation?

   Alarms should not activate during routine process variable changes, or from normal, expected cases of operation.

All of these criteria are routinely violated, as we will see in the following examples and later in the book. These examples are real-world, not theoretical!

4.1 Poor Practices

Example #1—Spare Pumps

This is one of the most common examples of mis-configuration of an alarm system. Consider a pump with a spare. In various valid and expected production situations, you may have:

- Both pumps off
- Either pump running by itself
- Both pumps running together for a long period of time
- Both pumps running for a short time as they are swapped

Throughout industry, the most common way that alarms are configured in this situation is the way that the DCS manufacturer made it "easy." Each pump has a digital run status point that shows if it is running or stopped. Each of these points is configured as follows:

> The "running" condition is declared the Normal state and the "stopped" condition is declared the Off-Normal state. Then, an alarm is often selected to sound when the pump is in the Off-Normal state, see Figure 4–1. (Configuration of the alarm is optional, and not required. But, usually the alarm is configured.)
>
> The result? Whenever you are intentionally running one pump or no pumps, you have one or two continuous standing alarms. This alarm configuration is very easy to implement, but does a really poor job of telling the operator that something abnormal has happened.
>
> The right way to think about this situation is as follows. An alarm should be produced only when the pump is not running *when it is supposed to be running*. This is a slightly more difficult alarm design to create, but is quite achievable. The same thought process must go into the creation of every alarm (see Figure 4–2).

So how do you do this? There are several ways that any control engineer could come up with if they use a bit of imagination. Here is a simple one:

> Who is it that knows, at any given time, how many pumps are supposed to be running? The operator, of course. Provide a little graphic element next to the pair of pumps that allows the operator to select 0, 1, or 2—the number of pumps that is *supposed to be running*. This is stored in the DCS. Then, configure a logic element in the DCS to compare the number of pumps running to the number desired! It can even incorporate a slight time delay before the alarm activates, specifically for when 1 pump is desired but 2 are running, as part of a normal pump swap.
>
> This design provides immediate notification to the operator when a pump fails, and produces no nuisance alarms. And all it takes is a bit of imagination and the utilization of capabilities that already exist in the DCS.

This is a good example because it illustrates several of the basic principles of good alarm management. The basic and most simple ability of the DCS,

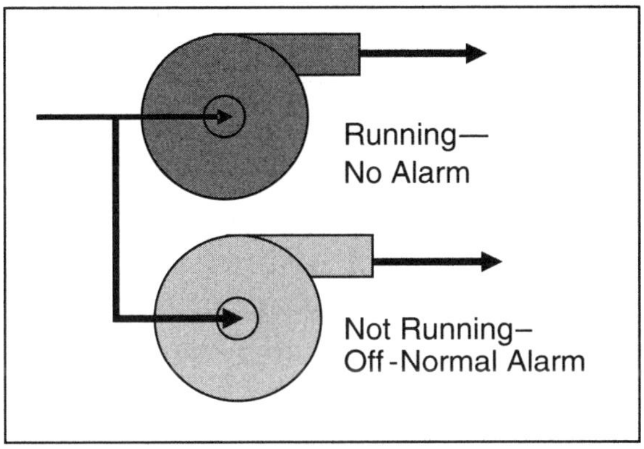

Figure 4–1 Spare pumps—alarming the wrong way.

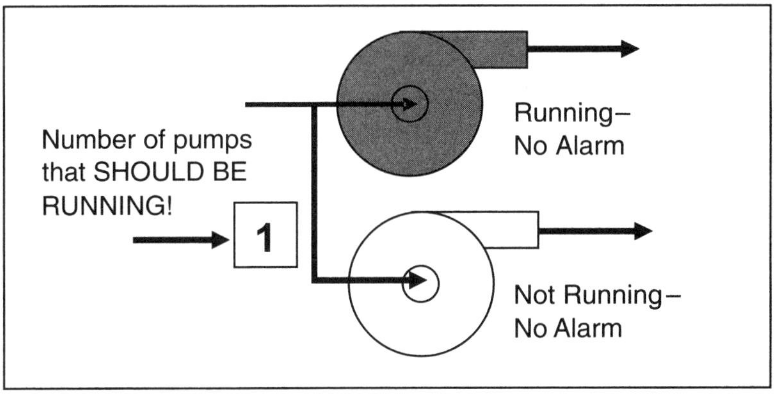

Figure 4–2 Spare pumps—alarming a better way.

if used without imagination, produces a poor result. Control engineers specify the alarms incorrectly, not because they are lazy, but because they are not aware of the principles of good alarm management. They follow the standard DCS configuration setups of alarms, thinking that the manufacturer would not have supplied them if they were not the right ones. But if you challenge a control engineer to make a situation better, you will usually get a good result.

Example #2—Shutdown Valve Position Switches

In this example, we see how common it is to configure alarms exactly backward from how they should be, again by using the "easy way out" in the initial configuration.

Consider an interlock that closes three different valves on a high pressure signal. This doesn't have to be from a complex Emergency Shutdown (ESD) System, it could be just an interlock within the DCS; the principle is the same. The interlock activation is an expected case of operation, otherwise the interlock wouldn't be configured.

Again, the control engineer uses the standard capabilities of the system without much thought to alarm management. Each valve has a position switch, and for each the "Open" position is declared "Normal," the closed position "Off-Normal," and an alarm assigned to that condition.

So: when high pressure occurs, and the interlock activates *as it is supposed to do*, then three alarms are generated when the valves move to the position *that they are supposed to move to*! The alarms say "Hey! I did what I was told!" This is exactly backwards. The operator has no action to take from these alarms, the system has responded correctly. (We are not talking here about an alarm relative to the high pressure itself, just these valve positions.) These useless three alarms act to obscure and interfere with the operator detection and response to valid alarms that will occur during this process upset.

The correct way to configure these alarms is, again, to look at the situation from the viewpoint of the operator and use a bit of imagination and logic. When the interlock activates, an alarm should occur *only on a valve that does not close*! The operator action is to ensure that the valve is manually closed or other isolation measures are taken. Again, logic must be used that takes into account the proper positions of the valves given the activation state of the interlock. This is not difficult to configure but it requires a control engineer have knowledge of these principles.

Example #3—The Automated Subsystem

Another very common alarm problem comes from the operation of automated subsystems. Often these are "skid-mounted," self-contained operations that cycle through a multi-step process. Dual filters that swap and backwash are a very simple example. For reasons outlined in the later section on alarms from external devices, the automated subsystem may come with many possible status indicators that can be connected to the DCS and alarmed.

We have often seen variations of the following. As the system cycles from step to step, an alarm is generated to notify the operator of that transition, as in:

> Alarm: "STEP ONE HAS COMPLETED."
>
> Then later, Alarm: "STEP TWO HAS COMPLETED."
>
> And so forth. We have seen hundreds of these alarms produced per week, telling the operator that things are working correctly and normally. We call these the "2-year-old toddler alarms" because what they are saying, over and over and over, is the same as *"Mommy! Daddy! Look at me! Look at me! Watch me!"*

The discussion with the control engineer responsible for the subsystem may go like this:

> *Know-it-all Alarm Management Consultant:* "The alarms from this system are configured backwards. An alarm should only be produced if it fails, as in "STEP 6 FAILED TO COMPLETE." Then the operator action is to troubleshoot and fix the situation. Otherwise, no alarm should occur when it works normally."
>
> *Engineer:* "But the operator really does need to know where this system is in the cycle!"
>
> *Me:* "Why? Does he have to input a weight or flow number or something to make it proceed from step to step?"
>
> *Engineer:* "No, it's automatic as long as it is working right. But they may need to know for a lot of reasons."
>
> *Me:* "OK, so why not build a simple graphic element to show the progress of the system in its cycle, the time that steps started or finished, the expected time of completion, and so forth. The operator can choose to look at the graphic whenever they really need to, when *they* choose. Create alarms like I mentioned earlier, that reflect an ABNORMAL, not a NORMAL condition. But don't use the alarm system in the way that is currently done, even if it is easy to do that way."
>
> *Engineer:* (aloud): "OK, I see what you are getting at, good idea." (Under his breath): "Smart-alec dweeb."
>
> *Me:* (note to self): Be more diplomatic. Well, *try*.

The alarm system is an *intentional interruption* to the operator. Here is an analogy as to why the alarm system must be reserved for abnormal situations only:

> *Heavy rain is pounding the cockpit windows as the airliner descends on final approach. While the pilot makes continuous power and course adjustments due to the high gusts, he reviews the overall situation. "Boy, I am earning my pay tonight. Zero visibility. Twenty degree course offset because of the strong crosswind. Speed plus 25 knots because of the gusts. Braking action on the runway reported poor by the last plane in. Engine 2 is surging a bit because of water ingestion. Breakout from the clouds will be right at minimums—200 feet. Approaching that now, get ready for the transition..."*
>
> *Just then, the cockpit door opens. The flight attendant reaches over, pulls the headphones away from the pilot's right ear, leans over and says loudly, "I thought you would want to know that the passengers all returned their seatbacks to the full upright position, just like I asked them to."*

Absurd? Yes—in an airliner—because people have made intelligent rules about proper cockpit practices and procedures. A plant operator, when managing process upsets, has many similarities to this pilot. Unfortunately, a typical DCS alarm system has many alarms similar in uselessness and negative, distractive effect as this interruption by the flight attendant!

Remember, the Alarm Summary Screen is not the control engineer's dumping ground for miscellaneous status information! In all of these cases, the important principles apply:

- Alarms must require an operator response
- Multiple alarms should not signify the same thing
- Alarms must activate only based on truly abnormal conditions, not expected cases of operation

CHAPTER 5

Creating an Alarm Philosophy

"Initiative is doing the right thing without being told to do so."

—Victor Hugo

The goal of the Alarm Management Philosophy document is to provide best practice guidelines for the definition, design, reengineering, implementation and ongoing maintenance of the alarm management system. Implementing these guidelines helps the operator take the correct action at the correct time in response to an abnormal situation. This is true if:

- Alarms are properly chosen and implemented.
- Alarms are relevant, clear, and easy to understand.
- Alarms are configured consistently in accordance with industry best practice guidelines.
- Alarms are presented at a rate that the operator can effectively handle.

- Operators can rapidly assess the location and relative importance of all process alarms.
- Operators can process alarm information during high frequency alarm actuation events.
- Alarm systems are properly controlled, monitored, and maintained.

The Philosophy document provides best practice guidelines for all aspects of alarm system design, implementation, and maintenance. It covers both new systems and modifications to existing systems. It is for both in-house use and contractor use during projects.

The Alarm Philosophy document provides a consistent and optimum basis for:

- Alarm Selection
- Priority Determination
- Alarm Configuration
- Alarm Handling Methods
- Alarm System Performance Monitoring
- Nuisance Alarm Resolution
- Alarm Detection, Presentation, and Annunciation
- Operator Interface for Alarms
- Operator Response to Alarms
- Alarm System Management of Change

The Alarm Philosophy is based on several key assumptions:

- No amount of alarm management will replace the constant surveillance of a qualified operator.
- Operators are trained on the alarm management strategy.
- Alarm Management enhances the operator's ability to make a judgment based on experience and skill.
- Operators will respond to all alarms, regardless of priority.

- Corollary 1: The system design therefore must not produce more alarms than the operator can respond to.
- Corollary 2: Alarms are never created solely upon the assumption that operators will fail to respond to a different alarm.
- Alarm priorities define the order of the operator's response.
- The alarm system is routinely maintained and evergreened.
- Alarm management includes all categories of alarms coming to an operator, including system alarms.

Certain portions of this book should be included in an Alarm Philosophy document in a modified form. The book is intended to provide extensive discussion and background about each topic, but an alarm philosophy need not be so lengthy. However, it is also important to not condense a philosophy too much. A major purpose of the philosophy document is to be used as guidance to people in project roles who are unfamiliar with the topic of alarm management. Enough detail must be included to give such groups the education they need so that their alarm implementations will be proper and consistent. A supplemental document such as this book, or a specific alarm management best practices training course, would help to prepare participants in an alarm improvement project.

To aid in this, an outline of an example Alarm Philosophy, reflecting best practices, is provided as an appendix to this book (See Appendix B).

CHAPTER 6

DCS Alarm Display Capabilities

"In a way, staring into a computer screen is like staring into an eclipse. It's brilliant and you don't realize the damage until its too late."

—Bruce Sterling

There are typically three methods by which alarms are displayed to a DCS console operator. They are:

- The Alarm Display functionality provided by the DCS manufacturer
- Custom graphics created by the owning company
- External "Lightbox" annunciators added to the DCS

We will individually discuss these capabilities.

6.1 The DCS Alarm Display

All DCSs come from the manufacturer with an Alarm Display. This is not referring to a dedicated piece of hardware; it is a pre-configured graphic that basically shows a scrolling list or multiple pages of alarms. Often, the operator selects one physical screen (CRT or LCD) from the several that they have available, and keeps this display up most or all of the time. The usual capabilities of these displays include:

- Sorting by alarm priority
- Sorting by chronological order
- Sorting by predetermined process area
- Color coding by priority
- Ability to temporarily "freeze" the display list during periods of high alarm actuation
- Ability to temporarily silence the alarm horn based on alarm priority
- Color choices and alarm symbology choices
- Displaying the measurement and the alarm trip point violated. Many will have a portion of this feature, and the best will have a live updating measurement value.
- Guiding the operator in responding to the alarm, by linking the alarm to the display used to control the measurement or system in alarm

Other capabilities may exist, depending on the manufacturer. Most are quite similar, but there are important differences. From an Alarm Management point of view, the important message is to thoroughly understand every option regarding this display that the DCS manufacturer gives you. Make explicit decisions about these options as you select them; do not just use the "default settings." Alas, this will involve the dreaded task of reading the system documentation.

If you are purchasing a new control system, be sure to make needed alarm display capabilities part of your specification. We still see many alarm system design omissions from the DCS manufacturers. Proper design should include the following elements:

- Priority systems that allow independent priorities per alarm
- Alarm summaries that update the alarm list or measurement values dynamically
- Ability to temporarily suppress the low priority alarm horn
- Navigation ability to go, in one click, from an alarm on the display to the proper graphic for diagnosing the relevant situation
- Temporary alarm scroll freeze to aid readability

The delivered systems will not improve without pressure from potential buyers.

6.2 Annunciated Alarm Priority

Alarm priority is a means to convey the seriousness of a specific process condition to the operator and drives the operator's responses. DCSs generally allow for multiple alarm priorities to distinguish alarms, as well as separate alarm priority assignment for each alarmable parameter of a point. Annunciated alarms are those that are communicated to the operator through an operator display and an audible notification.

The best practice principles of Alarm Management require that every individual alarm be assigned a priority using a logical and consistent approach. It is important that the DCS present alarms to the operator with a priority that has a consistent meaning. Note: this means that separate alarms on the same point should often have different priorities.

The best practice is to use *three* levels of annunciated DCS alarm priority. You may have a DCS that allows you to have many more than that. Do not succumb to the temptation of using them! Humans are wonderfully able to put things in three categories and to understand things that are in three categories. Four or five categories are about the maximum, more than that will get cognitively "blurred together" and become confusing rather than helpful.

Alarm systems from different DCSs may have differing nomenclature for priority levels. In this book, the levels of alarm priority will be designated as:

Critical[1]

Emergency

High

Low

Regarding the recommended three levels of priority (Emergency, High, and Low); there are three specific best practice recommendations.

- Every alarm priority should be assigned its own unique color. This generally involves reserving three colors solely for use in differentiating alarms. These colors shall not be used for displaying anything else on the DCS other than alarms of the matching priority.
- Every alarm priority chosen should have its own unique alarm sound. In a control room with several operating positions and consoles, this could pose a difficulty. If closely adjacent consoles have the same sounds, then the operator cannot use sound to detect a new alarm on their own console. The solution is that each console can use its own "family" of sounds for priority. It is also possible to use lights; we have seen consoles topped by a small stacked cylinder of three lights, with colors matching the alarm priority colors. These lights activate either instead of, or (better) along with the appropriate sound. In this way, if the sound volume is kept down, and if one operator is having a discussion with another operator at their console, the lights help to indicate the presence and location of a new alarm.
- It should be possible to turn off the horn for the lowest alarm priority during periods of high alarm loads. The operator doesn't want or need a continuing distraction from the low priority alarm horn during a major upset. Visual notification should remain in place. This practice must NOT be left in effect all the time. It should have a timeout feature after a few minutes.

We often see these basic principles violated. We see "alarm colors" used for all kinds of different graphic elements, and single alarm sounds assigned to multiple priorities. Even worse, we see priorities that have no sound at all—making it much less likely that an operator will initially see such an alarm. The result of such configuration decisions will be an alarm system that is less effective in helping the operator to properly detect, identify, and respond to alarms.

---

1. If you absolutely must have more than three categories, then use a fourth that we will call "Critical," which is above "Emergency" in the hierarchy. As you will later see, there are very few reasons to do this, but we will cover them. There will be very, very few "Critical" alarms. An additional possibility for a fourth priority is covered under "Alarms from Instrument Malfunctions."

It is well established through industrial studies, the EEMUA guidelines, and our own experiences from hundreds of projects that, for best results, alarm priority should be distributed as seen in Table 6–1.

These numbers are commonly referred to as the 80%–15%–5% distribution, and it is understood that the results for individual units may vary from these. Regardless of priority, all alarms require a response.

It is important to note that the alarm priority distribution guidelines shown above are the natural outcome of a sound and consistent alarm rationalization process, and not an artificially forced assignment simply to meet best practice guidelines.

Many DCSs have an alarm priority called "Journal," "Record-Only," or the equivalent. An alarm with such a Journal priority is recorded in the alarm event journal (i.e., historized), but is not annunciated in any way to the operator and does not appear in the Alarm Summary. Thus, in reality, such an "alarm" does not meet our criteria for the definition of an alarm at all. The Journal priority is often used to produce time-stamped records useful for determining the sequence of events in a post-event analysis. They are also sometimes used in verifying the proper activation of certain elements of an ESD or other automatic system. The analysis of so-called "Journal alarms" is covered later in the Baseline section.

6.3 Custom Graphics—Alarm-Related Guidelines

DCS operating graphic displays should act to always effectively help the operator control the process in the best possible way. Custom graphics are the most common method for conveying process information on a DCS. The proper design of such graphics is a book all by itself. We will summarize here just a few basic alarm-related principles. See Appendix A for a more lengthy discussion on effective operator graphics.

- **Keystrokes:** The DCS operator interface system shall be designed to minimize the number of keystrokes required to identify, verify, and assess an alarm. The system and graphics should be configured such that it is *never* necessary for the operator to type in a point name or graphic name.
- **Associated Graphic:** Every point with a configured alarm should have an associated graphic display on the DCS. This associated display should aid the operator in the proper diagnosis and mitigation of the event that caused the alarm. Methods by which the operator is quickly directed with a single keystroke or button-click

Table 6–1 Recommended Alarm Event Priority Distribution

Alarm Priority	Percentage of Total Alarms
Emergency	3–7% (5%)
High	15–25% (15%)
Low	70–80% (80%)

(i.e., one-touch access) to the associated display should be used. Many DCSs have this capability, but it must be configured.

- **Inherited Alarm Behavior:** Graphics should not be "hard-coded" with alarm behavior for points; the behavior should be consistent based on the configuration of a point's alarm and should change if the configuration changes. Example: If the practice is that an alarm is shown in a certain color based on priority, the graphic should detect the priority currently in effect on the alarm and display that correct color.
- **Alarm Status Indication:** A process graphic must visually and consistently highlight tags in alarm, whether or not the alarm is acknowledged, and the priority of the alarm. Alarms should always be the most prominent information or object on the display.
- **Colors:** Alarm colors are used ONLY to display alarms and never for anything else. If yellow is an alarm color, then yellow is never used as a text label, line color, border, or any other non-alarm-related element.
- **"Fat Finger" Contingencies:** Techniques should be used to minimize the possibility of operator mistakes, and provide validation and security measures. For example, a graphic element pushbutton that initiates an infrequent shutdown action should also require a step of confirmation of operator intention. Major process upsets have occurred by mistyping an input—for example, opening a slide valve to 47% instead of 4.7%! DCSs using membrane keyboards are particularly susceptible to this type of error.
- **Single Alarm Interface:** A single alarm interface should be used, namely that of the DCS. If alarms can come from sources nominally "outside" of the DCS, those should be brought into the DCS, if the DCS is used in any way to respond to the alarm. All alarms should be acknowledged only once; it should never be required to acknowledge the same alarm in more than one place.

6.4 Use of External Lightbox Annunciators

This might seem to be the logical place to discuss the proper use and effective integration of lightbox type alarm annunciators along with a DCS. But for reasons that will become very apparent, that full discussion will be delayed until after the section on Alarm Documentation and Rationalization. A full understanding of the proper way to create and prioritize alarms is needed prior to the discussion of lightbox annunciators.

CHAPTER 7

Baseline and Benchmarking of Alarm System Performance

*"If you torture data sufficiently,
it will confess to almost anything."*

—Fred Menger

An initial alarm system baseline and benchmark against industry best practices is essential to planning the improvement process. This is not difficult; a few simple and straightforward analyses will provide an excellent picture of the current performance level. A proper baseline should use at least eight weeks of continuous alarm system data. The data for each analysis must be based upon the alarms assigned to the span of control of a single operating position.

Good alarm analysis software should be able to perform all of the analyses in this chapter, and many others. It is possible to do these in a spreadsheet, although the data parsing and reduction will become tedious, speed is quite slow, and spreadsheet page size limits are easily exceeded when importing alarm journals. Frankly, using a spreadsheet to analyze alarm events is like using a water hose to fill an Olympic-sized swimming pool! The proper tool for alarm analysis is a real database.

All of the examples in this book are of real data, but slightly disguised to protect the embarrassed....

7.1 Alarm Event and Configuration Analysis

Alarms Per Day

The most important analysis is simple: Alarms Per Day (for a single operating position, as stated above). The number of alarms per day is a good indicator of the overall health of the alarm system.

This analysis, from a Honeywell system, shows two lines of alarm count data. The Annunciated alarms line is of all the alarms presented to the operator and is the most important. It is what was shown to the operator. The Honeywell system has an alarm suppression parameter that can override a point's alarm settings and prevent the operator from seeing any of the alarm events from that point. Instead, the alarm event goes only into the alarm journal (the operator never sees it), the same as if it had been initially configured as a "journal-only" alarm. This capability is often abused. If many points have this suppression override in effect, then far more alarms are being created than the operator is actually seeing. This is shown in the top line—Recorded alarms (see Figure 7–1).

For a thorough discussion of the issues and dangers around Alarm Suppression such as this, see the later section "Nuisance Alarms and the Dangers of Alarm Suppression" in the chapter entitled Alarm Handling Methods. In this case, about half of the alarm events produced from the alarm configuration are being suppressed from the operator's view. These alarms were initially intended to be visible to the operator. Alarm Suppression is a serious concern in this system.

Recommendations from EEMUA and other published studies are used to produce the two straight lines labeled as "Maximum Acceptable (300)" and "Manageable (150)." Alarm rates above around 300 alarms per day place the operator in the unenviable position of being forced to ignore

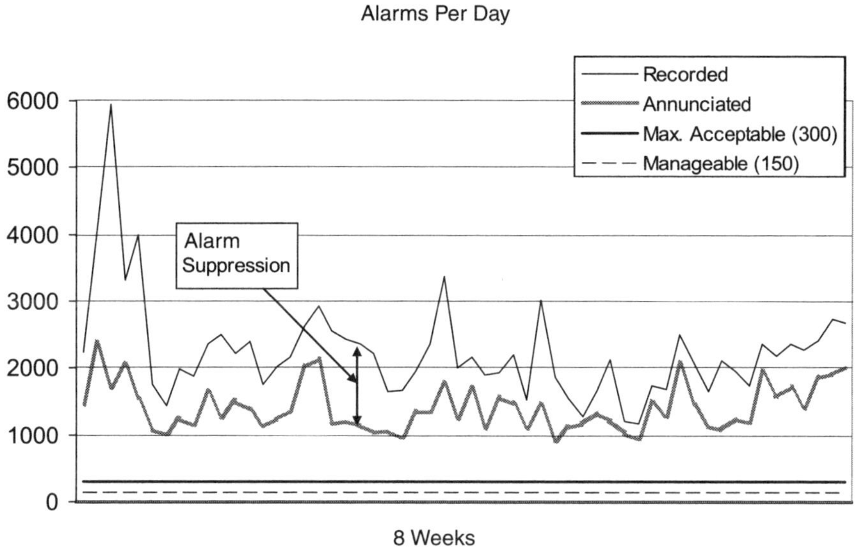

Figure 7–1 An example of an alarms per day graph.

many alarms—the number simply overwhelms their ability to analyze each one. In the above (and quite typical) data, this alarm system produces alarms at rates far beyond the operator's abilities of evaluation and response. Such an alarm system is not a useful tool to help the operator perform the right action at the right time! In fact, it is much more of a distraction or a hindrance to the operator.

Journal priority alarms, which are not annunciated (reported) to the operator, should not be a part of these analyses. As stated before, they are not really alarms at all and do not enter into the operator loading and response aspects of these analyses. However, an occasional separate look into their amount and rate of production should be made, because a DCS can actually be "slowed down" if it is generating and recording thousands of these "invisible events"—and this is not an unusual circumstance.

Most Frequent Alarms

The next analysis to perform is a simple ranking of the most frequent to least frequent alarm during the analysis period. The following chart and table are highly typical.

As is often the case, only 10 alarms are a significant fraction of the entire system alarm load, in this case 55%. (The analysis of hundreds of systems shows that the number is often over 80%, rarely is it less than 20%.) In fact, the top 4 alarms are over 40% of the load! Were they intentionally designed to annunciate so frequently? Of course not! Are they performing a useful function in their current configuration? Doubtful. The beauty of this analysis is that it can direct improvement efforts to where they will do the most good. Imagine finding the time and making the effort to improve only one alarm per week—to make it work as it was intended to work. In 4 weeks, this system would be improved by over 40%. Someone would be a hero (see Figure 7–2)!

In table form, the top 20 alarms are listed in Table 7–1 and account for over 60% of the entire system load. ("PV" is "Process value.")

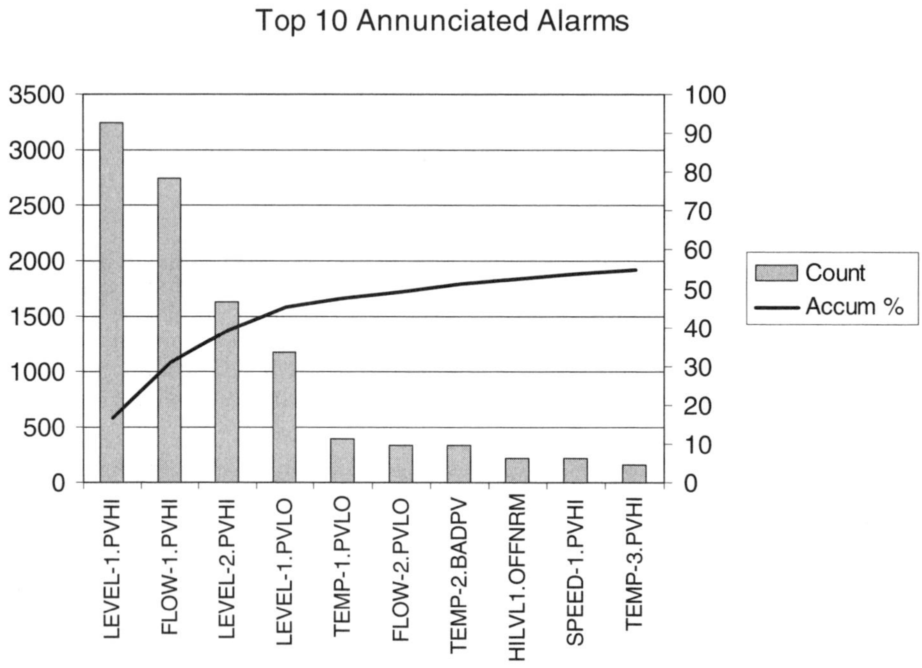

Figure 7–2 An example of a top 10 most frequent alarms chart.

Alarm Priority Distribution

Alarm priority is a means to convey the seriousness of a specific process condition to the operator, and is used to drive the order of operator's responses. Alarm priority must be consistently and logically assigned. The following analysis compares the priorities for the annunciated alarm events (not of configuration directly) during the analysis period, to well-known best practice ratios. Analyzing alarm events is superior to

Table 7–1 Example Table of Top 20 Most Frequent Alarms

Point	Alarm Type	Count	%	Cumul-%	Priority	Point Desc
LEVEL-1	PVHI	3248	16.65	16.65	H	C-301 BOTTOMS LEVEL
FLOW-1	PVHI	2744	14.19	30.84	H	CONDENSATE FLOW
LEVEL-2	PVHI	1624	8.32	39.16	H	C-201 LEVEL
LEVEL-1	PVLO	1176	6.05	45.2	H	C-301 BOTTOMS LEVEL
TEMP-1	PVLO	392	2.2	47.41	H	C501 COIL OUTLET
FLOW-2	PVLO	336	1.9	49.31	H	INLET FUEL FLOW
TEMP-2	BADPV	336	1.75	51.06	L	C701 SKIN TEMP
HILVL-1	OFFNORM	224	1.34	52.4	H	C310 LEVEL
SPEED-1	PVHI	224	1.34	53.74	H	WEST COMPRESSOR
TEMP-3	PVHI	168	1.03	54.77	H	C702 SKIN TEMP
HIGH-PRES-1	OFFNORM	168	0.98	55.75	E	MAIN FEED HIGH PRES TRIP
TEMP-4	BADPV	168	0.96	56.71	H	E-511 TEMP
TEMP-5	BADPV	168	0.95	57.66	H	E-512 TEMP
FLOW-3	PVHI	168	0.92	58.59	H	C401 OUTLET FLOW
VALVE-BYPASS-1	BYPASS-IN-EFFECT	112	0.57	59.16	L	MOV 332 BYPASSED
LEVEL-4	PVHI	56	0.55	59.71	H	SUPPLY TANK LEVEL
TEMP-6	BADPV	56	0.51	60.22	L	E612 TEMP
VALVE-BYPASS-2	BYPASS-IN-EFFECT	56	0.51	60.73	L	MOV 333 BYPASSED
FLOW-4	PVLO	56	0.51	61.24	H	C225 BOTTOMS FLOW
VALVE-BYPASS-3	BYPASS-IN-EFFECT	56	0.44	61.68	L	MOV 334 BYPASSED

analyzing configuration, for the purpose of identifying a priority distribution problem (see Figure 7–3).

In this example of another common situation, the use of High priority is far greater than best practice amounts. This system is significantly misconfigured, in a manner that is very common. Alarm Documentation and Rationalization (D&R) is the best solution to correct improper alarm priority configuration.

Chattering Alarms

Often, the most frequent alarm list contains several chattering alarms. A good working definition of a chattering alarm is one that comes in and clears at least three times in one minute. Obviously, such alarms are not being cleared by the operator taking action and adjusting the process! A list of chattering alarms should be produced and then acted upon. No alarm was intentionally designed to chatter, and they can all be fixed!

For a thorough discussion of the solution to such nuisance alarms, see the chapter "Common Alarm Problems and How to Solve Them."

Alarms Per 10 Minutes

"Burst rates" of alarms are quite important. Looking at alarms in 10-minute time slices gives a better picture of this than the daily amounts (see Figure 7–4).

An alarm rate of 10 or more alarms in 10 minutes defines the beginning of an alarm flood. In Figure 7–2, rates often exceed 20, 40, 60, or more alarms in 10 minutes. Such rates can continue for hours. During such periods, the likelihood of an operator missing an important alarm increases, as has been shown many times in the analysis of major accidents.

Alarm Floods

A refinement of the Alarms per 10 Minutes analysis is the Alarm Flood analysis. An Alarm Flood can last for many hours and include hundreds, or thousands, of alarm events. Alarm floods can make a difficult process situation much worse. A good default definition of an alarm flood is that it begins when the alarm rate exceeds 10 or more alarms occurring in 10 minutes, and ends when the rate drops below 5 alarms in 10 minutes. This analysis depicts the alarm floods occurring during the 8-week analy-

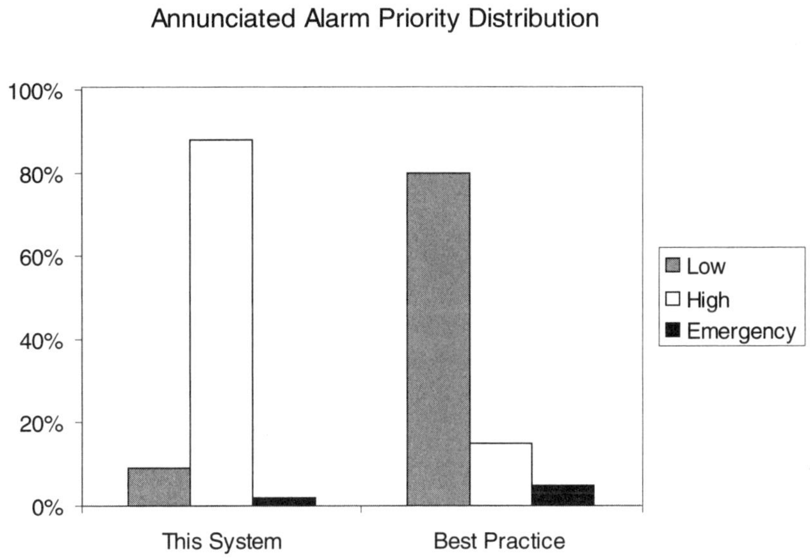

Figure 7–3 Actual vs. best practice alarm event priority distribution.

sis period, showing a breakdown by alarm count during the flood. Only alarms presented to the operator are considered (see Figure 7–5).

Alarm floods are a significant problem for this system. Most alarms produced by the system are during flood periods. Flood magnitude is very high, generally hundreds of alarms contained in each flood. There are over 14 floods per day on average (see Table 7–2).

Stale Alarms

Stale (or "Long Standing") alarms are in the alarm state continuously for more than 24 hours. Following their initial appearance, stale alarms provide no valuable information to the operators. They clutter the alarm displays and interfere with the operator's ability to detect and respond to new and meaningful alarms. The most common reason for a stale alarm is a lack of imagination in its initial creation and configuration. Most stale alarms are not indicating a truly abnormal situation.

Alarm Configuration Analysis

The preceding analyses will point you directly to the important problems to solve. They will let you use your limited time and resources to make the maximum possible improvement. Note that all of these analyses have been based on "dynamic data" (analysis of the alarm events) and not the

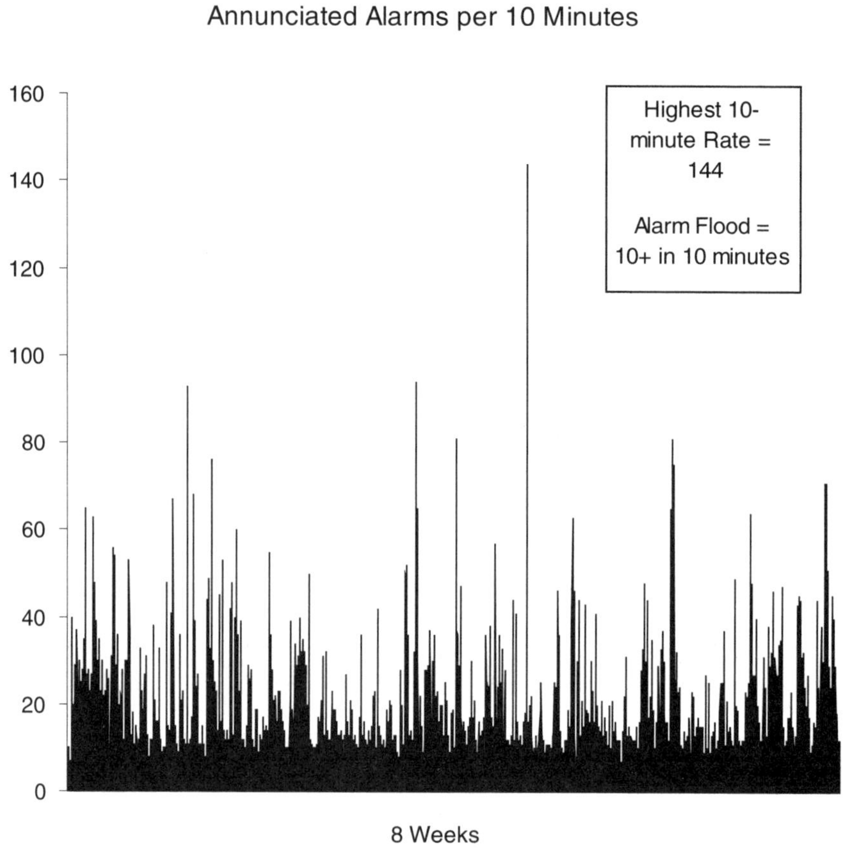

Figure 7–4 Example graph of annunciated alarms per 10 minutes.

underlying configuration of the alarm system. The EEMUA 191 document provides guidelines relating to configuration—configured alarms per control actuator, per analog measurement, and per digital measurement. Since the EEMUA 191 document was published, terabytes of information have been gathered and analyzed. The data conclusively shows that these configuration guidelines are not a valid predictor of the performance of an alarm system. Systems configured well within the guidelines can and do have *terrible* dynamic performance.

A note of warning. Alarm analysis software will generally have the capability to analyze alarms "by shift"—to compare alarms between shift-based operating teams. If you are interested in improving your alarm system, do not start by making use of these analyses! All you will accomplish is to alienate your operators; they will feel that the purpose of alarm

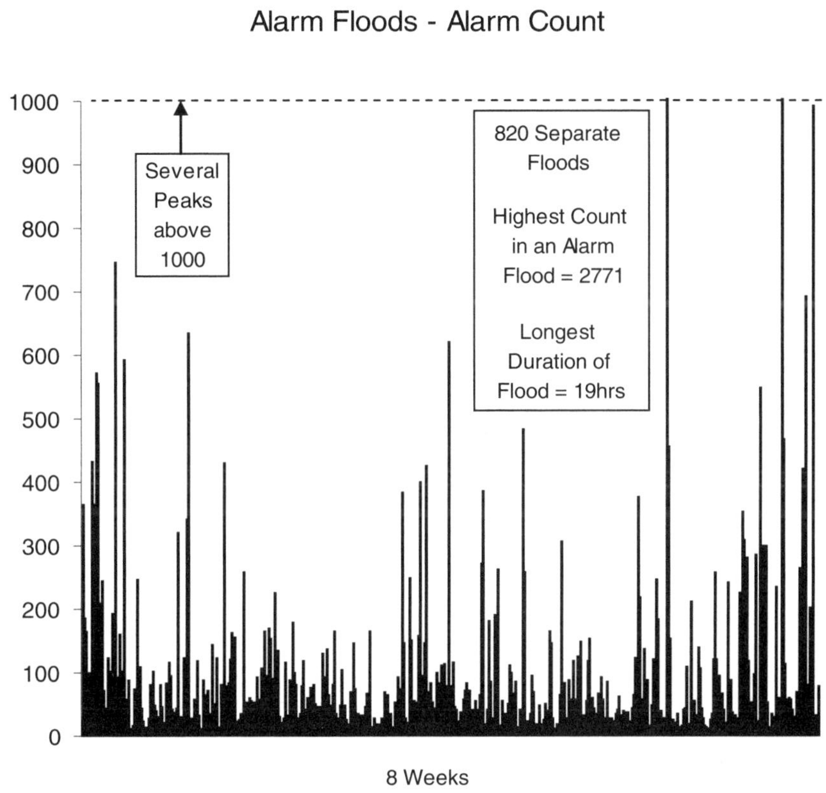

Figure 7–5 Example graph of alarm floods showing alarm count.

Table 7–2 Example Table of Alarm Floods

Number of Floods	820
Average Floods Per Day	14.6
Total Alarms in All Floods	58,376
Average Alarms per Flood	71
Highest Alarm Count in a Flood	2,771
Percentage of Alarms in Floods vs. All Annunciated Alarms	73.8%
Total Duration of Floods, in Hours	676.5
Percentage of Time Alarm System Is in a Flood Condition	50.3%

analysis is to somehow "get them." If an alarm system is performing poorly, there is no benefit to be had from these shift-specific analyses. It is not the operator's fault that hundreds of alarms chatter and go stale. The story changes slightly once a system is reengineered to an optimum performance. Then, shift-specific analyses may help identify potential operator performance or training issues.

You may also find it interesting (if you have time) to compare alarm rates during the day vs. the night. Usually, the impact of daytime operations and maintenance is clearly reflected in the alarm rate.

7.2 Alarm System Performance Levels

If desired, the results of the above analyses can be used to assign an alarm system performance level. Donald Campbell Brown, of the BP Upstream Technology Group, proposed a five-level scale using the nomenclature below. The category text descriptions are a mixture of his original wording plus some other thoughts.

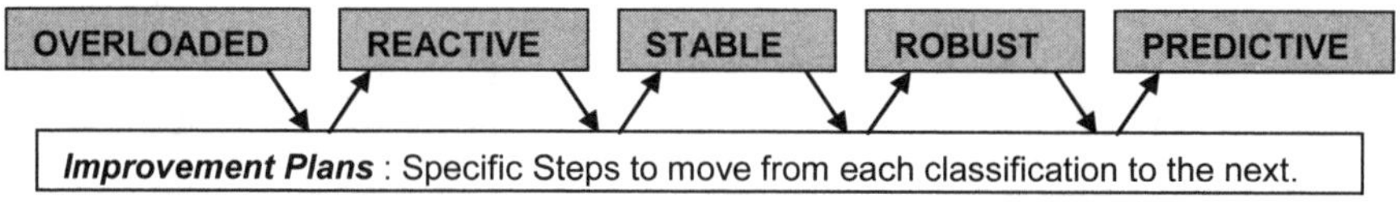

Overloaded

The alarm system is of little use to the operator and is often a hindrance rather than a support tool. This performance level is typically characterized by the following:

- A continuously high rate of alarms, with rapid performance deterioration during process upsets.
- The alarm system is difficult to use during normal operation and in practice ignored during plant upsets.
- Low operator confidence in the alarm system, which is often ignored for long periods.
- Important alarms are difficult or impossible to discriminate from less important ones, and alarm system gives little or no advance warning of plant upsets.
- Many alarms are meaningless or of little value.
- Alarms are often suppressed by the operator because they represent a nuisance, and are frequently then forgotten (i.e., never unsup-

pressed). Documentation and control of alarm suppression is unreliable.

Reactive

Some improvement compared to Overloaded, but the peak rate during upset is still unmanageable. The alarm system is still an unhelpful distraction to the operator for much of the time. Typically characterized by the following:

- Alarm system is more stable and useful during normal operation, but is often unusable in practice during plant upsets.
- The operator reacts more to the rate of alarm generation rather than to the detail of the alarms themselves.
- Alarm prioritization is known to be unreliable, but of some use.
- Alarm system gives some early warning of plant upsets.
- Some alarms are still meaningless or of little value contributing to overall noise level.
- Alarm suppression and general alarm system MOC are improved but may still not be under systematic control.

Stable

A system well defined for normal operation, but less useful during plant upsets. Compared to Reactive, there are improvements in both the average alarm and peak alarm rates. "Bad Actors" are resolved and under systematic control. Problems remain with the burst alarm rate. The alarm system functions well for normal operations, but is less useful during plant upsets. Typically characterized by the following:

- Alarm system is reliable during normal operation, providing early warning of impending plant upsets, but is less useful during a plant upset.
- Operators are confident in the appropriateness of the alarm prioritization, and react consistently and quickly based on priority.
- All alarms are meaningful and have a defined response.
- Alarm suppression and MOC are fully controlled.

Robust

Average and the peak alarm rates are under control for foreseeable plant operating scenarios. Dynamic and state-based techniques are used to improve the real-time performance. Typically characterized by the following:

- Alarm system is reliable during all plant modes, including normal operation and plant upsets.
- Operators have a high degree of confidence in the alarm system, and have time to detect and understand all alarms.
- Alarm system configuration is not subject to inadvertent change.

Predictive

Breakthrough performance is achieved on both the average and the peak alarm rates; the system fully meets all of the aspirations of the guidelines contained in Engineering Equipment and Materials Users Association (EEMUA) publication 191. Typically characterized by the following:

- Alarm system is stable at all times and provides the operator with the right information at the right time—in order to avoid process upsets or minimize the impact of any upset that does occur.
- Operators actively "patrol" the process schematics and corrects deviations before they are significant enough to cause an alarm.

Many companies are working towards achieving the "Robust" performance level. This level is reasonably achievable using existing technology at acceptable costs. However, if you have several Overloaded or Reactive systems, you are better off getting them all to Stable as quickly as possible, instead of spending time and resources moving other Stable systems to Robust. Achievement of this level of performance will require specific improvement plans to be adopted and implemented. This includes many of the items and strategies identified in an Alarm Philosophy Document, such as ongoing performance analysis, bad actor resolution, D&R, alarm change control, alarm shelving, and so forth.

7.3 Specific Alarm System Improvement Plans

From Overloaded to Reactive

- Alarm Philosophy Document Development: This is essential to improving an existing system. The document contains rules for consistent alarm creation, prioritization, handling, and modification. KPIs and periodic reports for Alarm System performance are set, roles and responsibilities are identified.
- Management of Change: MOC practices for alarms must be fully implemented. Alarm suppression, priority, and trip point changes must be controlled and authorized to ensure the overall integrity of the alarm system. Modifications must be clearly communicated to all affected personnel.
- Resolution of Bad Actors and Nuisance Alarms: A relatively small number of tags produce large numbers of alarm events. Identify and address the existing nuisance alarms (chattering, stale, and duplicate alarms) on the system. With a concerted effort many of the nuisance alarms in the system can be addressed immediately. This will provide some relief to the operators.
- Alarm Analysis: Alarm analysis is used in an on-going program to detect problems on individual alarms, categories of alarms, and sections of the process. Analysis is then used to drive definition and prioritization of the corrective action.
- Review alarm representation within operating graphics. Ensure the implementation of good principles of Human Machine Interface (HMI) for alarm systems.

From Reactive to Stable

- Alarm Documentation and Rationalization (D&R): This is the disciplined review of alarms within a process operating area relative to the operational objective of the alarm, the expected response to the alarm, and the conformance of the alarm to design standards. During D&R, alarm setpoints and priorities are determined using a consistent, engineered approach for each defined operating state of the equipment, in alignment with the Alarm Philosophy.

- Analysis and Publishing of Alarm Metrics: Metrics are used to produce on-going actions and improvement.
- Alarm Shelving: Alarm Shelving allows operators to temporarily shelve all or individual alarms of a tag. Once alarms are shelved, their status is tracked and they are periodically presented for enablement with operator confirmation. All alarms must be enabled unless identified as part of a defined, implemented, and controlled shelving strategy. Shelving is a short-term practice for dealing with instrument malfunctions, not a long-term "parking place" for troublesome alarms.
- Alarm Grouping: "Common Alarms" are provided for appropriate groups of related tags to prevent multiple alarms.
- Master Alarm Database: This is created during the D&R effort. The Master Alarm Database contains proper setpoints, priorities, causes, consequences, and corrective actions associated with each alarm.
- Implement automated Audit and Enforcement of on-line alarm configuration and settings to the Master Alarm Database.

From Stable to Robust

- Implement State-based Alarm Management: Alarms are minimized based on certain plant conditions (states) by modifying various alarm parameters dynamically. Alarm priorities, trip settings, and overall point alarm enable states are adjusted to ensure valid alarms are provided to the operator during all process conditions.
- Implement Alarm Flood Suppression: This consists of configured groups of alarms that can be suppressed in the event a pre-defined condition is detected. For example, if a compressor trips off, many of the resulting diagnostic alarms associated with the compressor should be suppressed, so that the operator can concentrate on minimizing the process upset that will occur from the loss of the compressor.
- On-line Alarm Response Manual: Making the Master Alarm Database data available on-line for operators.
- On-line Loop Performance Analysis: Continuous evaluation of the effectiveness of regulatory control loops to reject disturbances, with on-going bad actor reporting for maintenance action to correct tuning and/or valve performance issues.

From Robust to Predictive

Technology to fully achieve the Predictive performance level is still experimental and "bleeding edge." We discuss it here so you will know what to anticipate in the future. Predictive performance will involve the following kinds of technologies:

- Early Fault Detection: The early detection of the deviation of the process/equipment from its normal operation by monitoring a set of process variables. Deviations can be detected even when all the process variables are individually within their operating and alarm limits.
- Early Fault Diagnosis and Advice: The operator is advised with specific actions needed to prevent the upset from occurring.
- Procedural Automation: The automation of standard and emergency operating procedures associated with both normal plant transitions (e.g., start-up, shutdown, product change, and so forth) as well as critical corrective actions to recurring disturbances. Procedural automation allows for both on-line sequence execution and monitoring of steps in the procedure as well as collaboration between field and control room board operators.
- Extensive uses of operator support systems involving pattern matching, adaptive graphics, artificial intelligence, and other new experimental technologies.

Note: You can be sure that none of the items in this category will be cheap! But, like Advanced Process Control, when more fully developed and productized, adopters will acquire a significant operational and competitive advantage.

7.4 Analysis of Operator Actions

Would you like to easily obtain a competitive advantage? Would you like to make substantial improvements of your control system with little effort? If so, the same technology that collects and analyzes alarm events can provide you some remarkable insight into your control systems performance, since operator change events are also usually captured and available.

This is not referring to post-incident analysis, where you determine all of the alarms, actions, and events that occurred leading up to some specific incident. Yes, analysis software will aid significantly in such investigations but that is not the focus here. Consider the following analyses:

Controller Mode Changes

An event record is produced whenever a controller's mode (Manual, Automatic, Cascade, and so forth) is changed by the operator. These events are analyzed. Consider Table 7–3, which ranks the controllers that experience these changes.

Several interesting things are shown here. First, there are three controllers that experienced more than 100 mode changes in the week's data. Is it possible that they were designed to be operated in such a way? Unlikely. In fact, imagine asking your best control engineer to "design a controller where the correct thing for the operator to do with it is to change its mode about 100 times a week." You would get a funny look in return. But analysis will usually find controllers like this.

Operators do not change controller mode without a reason. They do not do it just for fun or to occupy their time; they perceive (rightly or wrongly) a *need* to do so. Therefore these controllers are not operating as designed. They need to be investigated and fixed. Otherwise you have just wasted the investment of installing them in the first place.

The analysis also shows controllers that are in MANUAL mode for much of the time. In reality, many controllers (for a variety of issues) run better in manual than in automatic! This is also unintentional on the part of the control engineer, and points out another opportunity for improvement. Similarly, controllers that are operated in modes other than their defined "Normal" mode are shown, and are opportunities as well. Many of the configured cascade control schemes in this example are apparently not working well.

Controller Setpoint Changes

Similarly, you can rank controllers whose setpoints are most frequently manually changed by operators. You will usually find some that are changed hundreds of times per week. This is obviously not an intentional design, and identifies loops that are candidates for a variety of improvements, such as (working!) cascades.

Table 7–3 Example Table of Controller Mode Changes in One Week

Point	Normal Mode	Change Count	% of time in Normal Mode	% of Time in MANUAL Mode
LIC 27	AUTO	111	96.5	3.5
FIC34	CAS	105	23.6	20.9
LIC 117	AUTO	101	74.8	25.2
PIC654	AUTO	78	99.8	0.2
FIC78	AUTO	74	0	1.5
LIC200	AUTO	70	99.3	0.7
FIC77	CAS	60	18.3	19.2
LIC01	AUTO	54	15.2	0.9
FIC288	CAS	54	20.7	3.5
TIC384	AUTO	49	59.7	31.7
LIC2088	AUTO	48	65.6	31.4
TIC309	AUTO	45	89.7	3.6
FIC897	CAS	45	7.2	44.8
FIX12	CAS	40	0	7.3
LIC611A	AUTO	38	95.8	0.3
FOC55	CAS	38	10.8	47
FIC22	CAS	37	0	0.7
FIC980	CAS	37	0	0.2
FIC400	CAS	36	32.7	21.9
FIC1000	CAS	34	26.6	31.1

Overall Operator Change Rate

In a similar manner, you can analyze changes in analog outputs (made when controllers are in Manual), and Digital Outputs (such as pump start/stops) and find candidates that stick out and cry for improvement. The above four items (mode, setpoint, analog output, and digital output) have something in common—they always have an effect on the process. While there are many changes that an operator can make and that may be logged by the DCS, not all of them affect the process (changing displays or access levels, for example). But since these four always do, a time-based graph of the sum of these actions is indicative of the overall amount of operator adjustment of the process. The next chart is an example.

In the chart, the overall operator change rate is often quite high. Stable, optimally performing systems have low operator change rates. All such changes disturb the system, whether for good or bad. Similar to alarm response, changes such as these should each be made by the operator deliberately and with full knowledge of the desired outcome. A change rate of more than 10 in 10 minutes is likely reflecting a period where the rush of changes may not all have adequate thought behind them. The higher rates shown (40 and above) are very questionable.

A variety of automation improvements can be used to reduce the amount of manual operator changes. It is straightforward to look into data during the peak change periods to see exactly what was being adjusted (see Figure 7–6).

Correlation of Operator Actions and Alarms

At this point in the book, we have seen graphs of alarm rates over time and of operator actions over time. Many of you engineers reading this are already screaming "CORRELATE THEM!" You are thinking, "Let's see if high alarm rates produce high periods of operator action, or vice versa. Let's try to find a causal relationship." You are going to be disappointed.

Start with a decent size database of a single system, containing operator changes and alarms. This data should span several months and contain at least a million events. Look to answer some of the following questions:

> Are periods of **high** operator action followed by periods of **high** alarm rates?
>
> Are periods of **high** operator action followed by periods of **low** alarm rates?

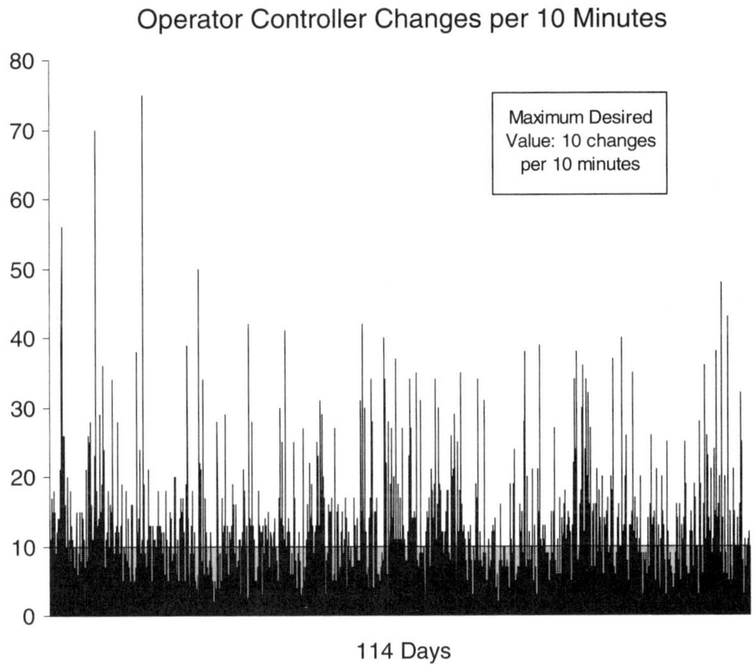

Figure 7–6 Chart of operator controller changes per 10 minutes.

Are periods of **low** operator action followed by periods of **high** alarm rates?

Are periods of **low** operator action followed by periods of **low** alarm rates?

Are periods of **high** alarm rates followed by periods of **high** operator action?

Are periods of **high** alarm rates followed by periods of **low** operator action?

Are periods of **low** alarm rates followed by periods of **high** operator action?

Are periods of **low** alarm rates followed by periods of **low** operator action?

And so on through the various permutations.

The answer to all of the above seems to be—wait for it—"Yes." You will likely find many examples of all of them being true. The same is true if you add in terms like "**no change in**" or "**medium**" for the alarm and action rates, and look for those situations in combination with "**high**" and "**low.**"

Basically you can "prove" anything if you pick the right spot in a large dataset, so really it seems nothing is proved. The answer is not found in general analysis of events in a "macro" sense; it requires examination of specific actions and alarms involving specific equipment in more of a "micro" sense—such as a detailed incident investigation.

Analysis of Changes That Require Management-of-Change Procedures

Certain changes made in the system should be done under Management-of-Change (MOC) control, as system safety and integrity can be compromised. Such changes must be properly evaluated, authorized and communicated to all affected personnel. Table 7–4 summarizes several such changes as recorded in a typical system. This can then be compared to the MOC records in an audit fashion to ensure that unauthorized, undocumented changes are not occurring.

There were almost 700 changes of types that need Management-of-Change (MOC) control, averaging about 6 per day. If we assume that all of these changes were authorized, note that the MOC system itself must be designed to accommodate the number of necessary changes, without an overburden of paperwork, and without compromising safety.

Usually, this analysis will greatly surprise people. In many systems, the staff has no idea of the quantity of important parameters that are being altered without proper analysis and authorization. DCS systems are notoriously easy to change and your operators know how to do it. Proper security and access control is a must.

Table 7–4 Example Table of System Changes Needing MOC Control

Summary of Changes in Alarms Needing Management of Change (MOC)	
Type of Change	**Change Quantity During Analysis Period**
Alarm Suppression	79
Alarm Trip Points	181
Alarm Priority	92
Point Range	121
Point Execution Status	105
Tuning Coefficients	0
Total	**685**
Average Per Day	**6**

CHAPTER 8

Monitoring and Reporting Alarm System Performance

"You cannot manage (improve or control) what you do not measure."

—Peter F. Drucker

Alarms, similar to process control points, collectively make up part of the DCS database configuration. Since the configuration of the alarm system is dynamic and subject to change, it requires ongoing maintenance to ensure accuracy, operability, and compliance with plant standards. Every control system should have a named Alarm Management System Champion, whose responsibility will be to maintain the integrity of the alarm system and ensure standards compliance. The role also includes maintaining the Master Alarm Database and reporting alarm system performance. If a person is not assigned to this role, sustained improvement is very unlikely.

Let's be very clear here. Without an ongoing commitment from line management, alarm improvement efforts will not get far. It is Operations that owns the alarm system, although staff groups like Engineering may have the most knowledge of it. Fixing an alarm system requires efforts from different groups, and the Alarm Champion must be backed up by sufficient authority to get things done. The Alarm Champion role is not a job to temporarily assign to an operator who has a temporary work restriction and has to "be on days."

8.1 Alarm System Key Performance Indicators (KPIs)

Measurement is fundamental to control and improvement, and improvement is best made against a pre-determined goal. The following alarm performance targets are achievable goals. Included is an interim set of targets specifically for systems that are beginning an improvement effort. Based on our experience, the chasm between the initial baseline of a system and the long-term targets may seem too wide to cross; the interim targets are an easier "first bite" to take as seen in Table 8–1.

8.2 Alarm Performance Reports

Alarm System performance should be periodically reported by the Alarm Management System Champion. The report should be distributed to managers, staff, and operators. In addition to the KPIs above, the report should also include:

- List of alarms which are not in service (e.g., shelved)
- Frequency analysis for most frequent alarms and chattering alarms, showing top 10
- List of long standing (stale) alarms
- Creation of a Nuisance Alarm List and progress made against the resolution of the alarms on that list
- Definition of action plans to improve performance compared to the KPIs, and progress of those plans

Alarm system performance should be monitored regularly (weekly) by the alarm system champion. Frequency of the alarm performance reports to others in the company may be tailored to the specific needs of the individuals' roles. For example, at one chemical company the site manager

Table 8–1 Recommended Alarm System Key Performance Indicators

Key Performance Indicator (KPI)	Interim Target for Systems Undergoing an Alarm Improvement Effort	Long-Term Target
Target Average Process Alarm Rate	<300 per day	< 150 per day
Percentage of time alarm rate exceeds Target Average Process Alarm rate	5%	0%
Alarm Event Priority Distribution based on at least one week of data	~80% Low, ~15% High, <=5% Emergency	~80% Low, ~15% High, <=5% Emergency
Suppressed Alarms	Zero (Unless as part of defined Shelving, Flood Suppression, or State-based Strategy)	Zero (Unless as part of defined Shelving, Flood Suppression, or State-based Strategy)
Chattering Alarms	10 occurrences or less in a one-week period	0 per day
Stale Alarms (more than 24 hours old)	20 or less in a one-week period	0 per day
Floods (10 to 20 alarms in a 10 minute period)	<= 5 per day	<= 3 / day
Floods (>20 alarms in a 10 minute period)	<= 3 per day	0 per day
Process changes in Alarm Priority, Alarm Trip Point, Alarm Suppression Status, Point Execution Status	None that are unauthorized. None that are not part of a defined Shelving, Flood Suppression, or State-based Strategy.	None that are unauthorized. None that are not part of a defined Shelving, Flood Suppression, or State-based Strategy.

reviews the performance of the alarm systems as part of the overall plant performance effectiveness once a month.

Commercially available alarm system performance monitoring and reporting software facilitates the gathering, analysis, and reporting of the metrics via a company intranet or e-mail.

It is advisable to create special reports after major upsets, or incidents such as spills, that detail the alarm system performance prior, during, and after the incident.

CHAPTER 9

Alarm Documentation and Rationalization (D&R)

"Finally the alarm system makes sense."

"The alarm system is useful now.
It sure wasn't before."

"You can understand the alarms now—
they have real meaning."

"This is the best thing we have ever done."

"The alarm system is now under control!"

"I'm not constantly dealing with a bunch of
incomprehensible alarms anymore."

—Quotes from operators
after alarm system improvement projects

In this section we will cover a powerful, rigorous, best practice alarm improvement method called Alarm Documentation and Rationalization, or "D&R."

9.1 D&R Overview

D&R is a sound, consistent, and logical methodology by which alarms are determined, prioritized, and documented. Alarms resulting from the methodology are said to be "rationalized," a very desirable state. D&R involves a thorough re-examination of every alarm configured on the system, and on other points for which it is possible to configure an alarm.

D&R is used in the following ways:

- To configure the correct alarms on an existing system. This invariably results in a reduction in the number of configured alarms, based upon hundreds of projects. It is not the specific intent of rationalization to reduce that number; the reduction is a natural consequence of the process.
- To correct a mis-configured system for performance improvement
- To initially determine the proper alarm configuration on a new system
- To ensure consistency in alarm settings
- To eliminate duplicate alarms
- To ensure proper and meaningful Priority and Trip Point settings
- To configure alarms on points added or modified by projects or as needed based on changes in process operations
- To provide detailed alarm information for use by the operators
- In conjunction with Process Hazard Analysis (PHA), Safety Interlock Level (SIL) revalidation, or Layer of Protection Analysis (LOPA), if alarms are specified
- To create proper configuration of nuisance alarms as they are identified
- To create the Master Alarm Database, used as a reference for operator training and use, and for State-Based alarm management, flood suppression, and Audit/Enforce mechanisms

The basic methodology of a D&R is simple. For each point on the system, a team of knowledgeable people do the following:

- Discuss each configured and possible alarm on that point

- Verify that any configured alarm should exist at all
- Verify that an alarm does not duplicate another similar alarm that occurs under the same conditions. If so, keep the one that best indicates the root cause of the abnormal condition.
- Determine the proper priority of each alarm in this fashion:
    - Determine the severity of the consequences that will occur if no action is taken in response to the alarm
    - Determine the time available for the operator to successfully respond to the alarm to avoid those consequences
    - The combination of these factors determines the alarm priority
- Document as much of the following as is practical:
    - Possible causes of the alarm
    - Method of alarm verification
    - Proper operator response to the alarm
    - Other points likely to be involved with the alarm
    - Relevant Operating Procedure, PHA, or other references
- Determine the proper trip point for the alarm based on:
    - Examination of process history
    - Examination of relevant operational procedures
    - Examination of equipment and safety system specifications
- Note any modifications to the alarm needed, such as introduction of logic, reconfiguration of alarm type, alarm message rewording, DCS graphic changes, and so forth
- For processes with different operating states, several different alarm values may be required. During D&R, record the appropriate settings for each state. See the section on state-based alarming.

D&R is not cheap. Most systems have thousands of points. As you can imagine from the previously mentioned lists, it can be a quite lengthy, disruptive, and expensive effort in terms of internal resource utilization. However, if you follow the methodology and suggestions covered in this chapter, you will find that it can be accomplished with relative efficiency.

For an existing system, D&R should be undertaken only after alarm "bad actor resolution" is completed. For that, see Chapter 13 "Common Alarm Problems and How to Solve Them." For a new system, D&R should be incorporated as part of the project's initial alarm configuration design, where it will be much more efficiently accomplished.

This chapter will cover all of the issues involved in performing a successful D&R. The various reference works on Alarm Management discuss several possible methods for accomplishing D&R. Some are quite academic and impractical. The method put forth here is proven through hundreds of thousands of successfully rationalized points.

9.2 D&R Participation and Reference Information

For proper D&R, it is a recognized best practice that participation includes:

- Operations Technicians (Operators). Two experienced operators are recommended, specifically from different shifts (or teams), to increase the acceptance of the major change in the alarm system that will result from performing the D&R
- Production and/or Process Engineers familiar with the process workings, economics, and with the DCS control system
- Safety and Environmental Engineers (part time as needed)
- Process Control Engineers (part time as needed, particularly when advanced control strategies or ESD logic are discussed)
- Maintenance Engineers (part time as needed, when specific equipment is discussed)

Other individuals with knowledge of the process unit, its operation and specific equipment, its advanced control schemes, unit hazards, and the alarm philosophy, will be needed periodically. The entire team must understand the alarm philosophy before starting the rationalization.

Information required for a thorough D&R includes:

- Unit P&IDs
- Operating procedures
- DCS configuration data for every point
- Results from HAZOP, PHA, LOPA, or similar reviews

- Emergency Shutdown System logic diagrams (and people knowledgeable about the logic)
- DCS graphic printouts
- On-line access to historical process data
- On-line access to historical alarm data

During a D&R, all alarmable DCS points should be examined, along with any other systems which provide alarm or abnormal situation notification to the board operator.

9.3 The Alarm Rationalization Grids

Areas of Impact and Severity of Consequences

To prepare for D&R, you must, in advance, build three simple grids. The first and most important of these is the grid for Areas of Impact and Severity of Consequences. Table 9–1 is a recommended example. To use the grid, each Impact Category is discussed separately. The discussion is "How severe are the consequences, if the alarm occurs and no operator action is taken in response?"

There are several important points to be made about this recommended example:

- Any single scenario may have different severities in the different impact categories, this is normal and expected. Assign the overall severity for the event to be whichever one is *worst*. You will not do yourself any favors by trying to come up with complicated point scoring systems. This overall impact is simply classified as Minor, Major, or Severe. (If it is "None", then you don't need the alarm!)
- Note that there are only three areas of impact and three severity levels. Remember, humans are wonderfully able to put things in three categories and to understand things that are in three categories. The Keep-It-Simple philosophy really applies here. You will not do yourself any favors by expanding either the Impact categories of Severity levels. The eventual output of this process is the assignment of one of three priorities, so a 10x10 grid will just waste your time.

Table 9–1 Areas of Impact and Severity of Consequences Grid

Impact Category	Severity: NONE	Severity: MINOR	Severity: MAJOR	Severity: SEVERE
Personnel Safety	No injury or health effect	Slight injury (first aid) or health effect No disability No lost time recordable	Injury affects work performance maximum one week Reversible health effects (such as skin irritation)	Lost time injury >1 week, or worker disabling, or severe injuries, or Life Threatening
Public or Environment	No effect	Local environmental effect Does not cross fence line Contained release Little, if any, clean up Negligible financial consequence Internal or routine reporting requirements only	Contamination causes some non-permanent damage Single complaint Single exceedance of statutory or prescribed limit Reporting required at the local or state agency level	Limited or extensive toxic release Crosses fence line. Impact involving the community Repeated exceedances Uncontained release of hazardous materials with major environmental impact and 3rd party impact. Extensive cleanup measures and financial consequences Reporting required at the state or federal agency level
Costs/ Production Loss/Downtime/Quality	No loss	Event costing <$10,000. Reporting required only at the department level	Event costing $10,000 to $100,000 Reporting required at the site level	Event costing >$100,000 Reporting required above the site level

- Do not try to separate out "quality" or "downtime" or "equipment damage" into totally separate areas of impact from "cost." Those things have costs associated with them that can be readily determined, and you will just be wasting your time trying to differentiate them.
- This grid is the core element of D&R and is not the place to be taciturn! Use lots of words and examples so that it is clear what types of consequences fall into the three severity levels. The purpose is that different groups of people evaluating the same event will come to the same severity assignments, and to do that you need clear words describing the boundaries of the assignments.
- You will find it easy to define your "Minor" category and your "Severe" category and may find it more difficult to define the middle or "Major" category. Actually, it is just everything in between, and falls out if you define the end points first.
- You may have similar looking grids from your PHA or LOPA process. Do NOT use them unmodified for D&R! They are specifically designed for risk assessment, which is a very different process.
- We have found that using "authority reporting level" as shown is a very effective differentiator of the categories. The analogy is: your teacher makes you stay after school (Minor), you are sent to the principal's office (Major), you have one phone call from the police station (Severe).
- The grid is NOT equating the dollars in the Cost category with the Environmental effects or Personnel Injury potentials in the other categories. Each category is evaluated separately and independently; they are shown in one grid for convenience. If your organization has very high political correctness issues with this, then separate them into three pages. You are not a bad person if you care about costs as well as safety and the environment! An alarm that exists solely for monetary reasons, of *any* priority, is a perfectly legitimate alarm to have—ask your shareholders! Never forget that the alarm system is there to help you control the process, and the process is there to make you money.
- During the D&R, begin discussing each alarm by discussing this grid, rather than discussing the alarm causes or other items. If there are no consequences from the failure to take operator action in response to the alarm, then you eliminate the alarm and move on.

Special Guidelines: Probability

It is inappropriate to consider "probability" in discussing alarms during D&R. The assumption is that the alarm (however improbable the process situation) *has occurred.* The consequence to be considered is the event that will take place if no operator action is taken in response to the alarm. Alarm Rationalization is not a PHA, or SIL, or LOPA review. Such probability assessment and risk analyses are used to determine equipment specifications and redundancy in a system, not the priority of an alarm when an event does happen. Those equipment decisions have already been made by the time you are doing D&R. The purpose of D&R is to assign the alarm's priority, not to discuss the need for redundant transmitters for this signal, or if the probability of occurrence is really less than one in 10,000 years. If the D&R discussion brings up significant issues or items that those studies may have overlooked, note them on an action list for referral and move on. Do not get sidetracked.

Special Guidelines: Multiple Failures

It is also inappropriate to assume multiple cascading failures in discussing an alarm consequence scenario. This is best explained by example.

> *Consider a vessel that has a high pressure alarm. The vessel has a pressure relief device that actuates somewhat above the high alarm setting and is routed to the flare.*

During rationalization, it will be assumed that all protective systems (e.g., ESDs, pressure relief devices, other independent alarms) are active and functional. If those devices are needed, we assume in D&R that their design and reliability are proper. The D&R is not an equipment design exercise.

Failure to respond to the high pressure alarm would therefore have Environmental (flaring) and/or Economic (loss of product to the flare) impacts, but no Personnel Safety impact. In terms of setting the appropriate alarm priority, it would *not* be appropriate to say that the consequence would be that in the high pressure scenario, the relief device would also fail, the vessel would rupture, and personnel could be injured (i.e., a Personnel Safety impact).

If you do not follow this principle, then your alarm priorities will come out vastly skewed to the high end and you will not accomplish your alarm improvement goals. In actual practice, most of your alarms will be defined from the Environmental or Cost consequences. In a modern process facility, there are actually very few cases where an alarm, followed by operator action, is the method by which harm to a person is avoided. The most common such cases where this is true (things like toxic and flammable gas detectors) are covered in the section on Specific Alarm Design Considerations.

(As an aside, we actually heard the following during a D&R: "But if this relief valve blows, it is pointed towards the walkway and someone could get exposed." The right answer is to reroute the discharge, NOT to increase the priority on the alarm!)

Maximum Time Available for Response and Correction

The next grid you must configure is simpler. "Maximum time to respond" is the time within which the operators can take action(s) to prevent or mitigate the undesired consequence(s) caused by an abnormal condition. This response time must include the action of outside personnel following direction from the console operator. It must also account for the time needed for the process to successfully respond to the change.

To clarify, this is *not* how long it actually takes the operator to take the action, which might only be a few seconds in the best possible scenario. It is how much time is *available* to take the effective action from when the alarm sounds to when the consequence becomes *unavoidable*, regardless of the action. Remember these examples?

> *The alarm sounds, indicating that the self-destruct of the starship Nostromo has just been activated. From now, you have 15 minutes to re-engage the cooling system or the core will explode. Re-engaging the cooling system only requires about 1 minute of work, but you have to make your way past the alien to get to the coolant controls!*

> *The biological seals of the underground laboratory have been breached. You have the special key to defeat the thermonuclear cleansing device. Inserting and turning the key only takes a few seconds, but you have to climb the central core and dodge the automatic sentry lasers to get to the lock!*

Most process plant operators do not have to dodge aliens or lasers to take corrective action. Their obstacles have more to do with distractions from mechanical problems, phone calls, maintenance interruptions, engineer or staff interruptions, or the activation of other distracting alarms. The operator's ability to respond to an alarm in a timely fashion determines the degree of success in preventing loss. The consequences of an uncorrected alarm generally worsen with the passage of time.

During an abnormal condition, the board operator is confronted with making decisions on numerous tasks that must be performed in an appropriate sequence. The timing and the order of executing these tasks determines the outcome of the operator's effort. For example, if two process variables are deviating from normal and can potentially cause the same significant loss, the operator must quickly decide which variable to address first. In such a case, the operator must take action to address the variable that is more volatile or can reach the point of loss in the shortest time.

Therefore, the shorter the time available to respond, the higher the priority of the alarm will be, assuming equal consequences can result.

We will follow the Keep-It-Simple principles and use only three meaningful categories for Maximum time to respond. They are:

- "Less than 3 minutes" or "Immediately." Drop what you are doing and respond to this alarm right now. Hang up the phone. Interrupt the long-winded engineer talking about temperature profiles. Tell the maintenance technician that you will look at his permit later but you can't do it now.
- "3 to 10 minutes" or "Rapidly." Quickly wrap up the phone call to the lab about the last sample. Tell the maintenance tech where the analyzer room is. But don't start anything new, this alarm has to be dealt with.
- "10 to 30 minutes" or "Promptly." This alarm must be dealt with, but you can first call the field pumper to start that transfer. Or, you can finish talking to the engineer about next week's test. But not both.

Most companies choose one of the two options in Table 9–2, and it makes little difference which is picked.

There are some important points to be made about either of the recommended examples above.

- Do not be concerned during the D&R with trying to precisely determine if the time available is 9 minutes or 11 minutes. The operators in the meeting will easily be able to put the scenario into one of the three categories, engineering calculations are not necessary!

Table 9–2 Maximum Time Available for Response and Correction Grid

"Text Version"	Classes for Maximum Time To Respond—Option 1	Classes for Maximum Time To Respond—Option 2
<No Alarm>	> 30 minutes	> 30 minutes
Promptly	10 to 30 minutes	15 to 30 minutes
Rapidly	3 to 10 minutes	5 to 15 minutes
Immediately	< 3 minutes	< 5 minutes

- Note the upper limit. If a condition does not require a response within 30 minutes, then the condition does *not* meet the criteria for the creation of an alarm. While an operator may have a "time horizon" of several hours or more in adjusting process parameters and monitoring their effects, it is *inappropriate* to sound an alarm for which no action is required for more than 30 minutes. Alarms are to signal conditions that require quick action and must have a characteristic of urgency. Something that can be avoided for more than a half hour with no effect is not an event requiring quick action.

This is not an absolute principle; there will be exceptions. An example of an exception is a reactive monomer tank, where a rise in temperature may not produce adverse consequences for many hours, yet the consequences are severe and action must begin to resolve the problem as soon as the rise is detected.

Another exception example would be an alarm of the failure of a system that acts to protect the long-term health of equipment—such as a corrosion inhibitor addition system. Failure to take action on the alarm might not have consequences for weeks or months, but the system is needed and the failure must be addressed, and not forgotten about. The general rule is that response to such an alarm should be the initiation of a maintenance request before the end of the shift. The need for the alarm system to retain a sense of urgency allows for such exceptions as these.

Priority Determination Grid

The third (and final!) grid puts together the results of the first two and determines the most appropriate priority for the alarm. See Table 9–3 for the recommendation.

There are several important points to be made about this recommended example.

- Note that every Emergency alarm requires immediate action on the part of the operator.
- Experience has shown that following this methodology will result in an alarm priority distribution close to the best practice recommendations of ~80% Low, ~15% High, and ~5% Emergency.
- The grid system may produce a priority that someone "just can't live with." Sometimes this is because of a prior incident, such as a spill, and "By gosh, we will have an Emergency alarm on that thing so there can be no excuse in the future!" In such a case, it is acceptable to override the priority recommendation from the grid and choose another, provided that the reason for such an override is written down as part of the alarm documentation.
- It is recommended that every Emergency alarm have a pre-alarm set to High priority, if practical and if there is adequate time for the operator to take effective action in response to the pre-alarm. On such things as toxic and flammable gas detectors, this is usually not practical and the initial alarm is Emergency priority. (See the chapter "Specific Alarm Design Considerations.")

Table 9–3 Severity of Consequences and Time to Respond Grid for Alarm Priority Determination

Maximum Time To Respond	Consequence Severity: MINOR	Consequence Severity: MAJOR	Consequence Severity: SEVERE
> 30 Minutes	No Alarm[a]	No Alarm[a]	No Alarm[a]
10 to 30 minutes	Low Priority	Low	High
3 to 10 minutes	Low	High	High
< 3 minutes	High	Emergency	Emergency

a. In actuality, the >30 minute situation is not to be ignored ("No Alarm"), instead the alarm should be re-engineered so that it has the needed characteristic of urgency.

Computerized Alarm D&R software makes doing the rationalization described above a simple matter of checking on-screen boxes during the discussion of each alarm. You can do without such software, using spreadsheets for example, but your productivity will be much less.

The Fourth Priority—"Critical"

You may have an internal need or requirement to define a fourth, highest-level priority. This would be for the "most critical" of all of your alarms. From a rationalization perspective, any Critical priority alarms should first meet all of the requirements to be rationalized as an Emergency priority per the method above. Then you must develop and follow clear, explicit, and consistent rules by which a decision is made to "promote" a specific Emergency priority alarm to Critical priority. The list of Critical alarms should be quite small—on the order of a dozen or less per operator position.

Alarms Specified by Process Hazard Analysis (PHA) Teams

In many plants, periodic safety reviews of facilities are required. It is exceedingly common that the team convened for this task will come up with dozens of new alarms to be implemented. The team is usually unknowledgeable about alarm management, or alarm philosophy, or perhaps even other alarms that already exist. Whenever such a team is convened, it is important that they are given the alarm philosophy, including specifically the requirements for D&R. The team must be required to perform D&R on any new alarm they recommend. This will help reduce the tendency of such teams to "throw an alarm at the problem."

It is important during D&R to not delete any alarm that is there as a result of a prior PHA effort, even if no one in the room can figure out why it exists or even if it duplicates some other alarm. There are usually stringent MOC policies that must be followed in changing such alarms, often requiring the approval of the prior PHA team.

Safety Interlock Level (SIL) Alarms

For identical reasons around MOC, it is also important during D&R to not delete any alarm that is there as a result of a SIL design. However, some engineers believe that any alarm generated as an output from a SIL design must automatically be assigned the highest possible priority. Not so. "SIL alarms" should be rationalized in the same manner as every other alarm and may well come out as Low priority. If your facility is not heavily involved with SIL analysis, equipment, and alarms, then consider yourself lucky! If it is, there are other books and standards that you should become familiar with.

The later chapter "The Death of the Lightbox" covers many of the issues involved around "Critical" priority and also "SIL Alarms."

9.4 Alarm Documentation

All rationalized process alarms should also be documented. The documentation should include all information required to define the alarm, its purpose, and the data used in rationalization. For new projects and incremental changes to the unit, alarm documentation and rationalization should be provided as part of the project scope, accompanying any other required project documentation (i.e., Management of Change documents).

The following items should be documented for each alarm:

- Possible causes of the alarm
- Operator response or recommended corrective actions for the alarm
- Potential consequences if there is no operator response to the alarm
- Time available for operator to respond and mitigate identified consequences
- Any reasons for over-riding priority recommendations determined by the rationalization principles
- Any modifications needed to the alarm, such as introduction of logic, reconfiguration of alarm type, alarm message rewording, DCS graphic changes, building of new common alarms, and so forth

Optional items (time permitting) to document include

- Method of alarm verification
- Other points likely to be involved with the alarm
- Relevant Operating Procedure, PHA, or other references
- Relevant P&ID number or hazard scenario number

The output of the D&R is a Master Alarm Database. Operations should have electronic access to this database (paper printouts are hopeless).

Methods of making the database available through the company intranet via a web browser are common. There are also methods to make the relevant alarm data available via a single click call-up on the DCS graphic where the alarm is displayed.

The Master Alarm Database has several other important uses, particularly for alarm auditing and enforcement, that will be covered later.

9.5 Alarm Trip Point Determination

Alarm trip point settings should cause alarms to actuate when there is a need to alert the operator to conditions or events. Alarm trip points should be selected to provide adequate response time to plant operations. Care must be taken to ensure alarms on analog signals are not duplicated by independent digital alarms provided by the ESD system or other interfaces, or vice versa.

Alarm trip point settings must be chosen to be consistent and cooperative with limitations originating from several related areas. Examples of these related areas include:

- Environmental
- Equipment reliability
- Safety systems
- Product quality
- Process design
- Personnel safety
- Economics
- Process history
- Plant dynamics
- Proximity to ESD settings
- Instrumentation
- Equipment design

The conference room where the D&R is conducted should have computer access to process history. This becomes essential when verifying the proper trip point setting for analog alarms. Alarms should not be set right in the middle of "where the process likes to live." Any measurement uncertainty should be considered as well. As shown in Figure 9–1, statistical analysis methods can be used to compare alarm settings to process history. This can determine optimal alarm settings that truly indicate the excursion of the process into abnormal territory.

In Figure 9–1, raising the trip point from 270 to 274 reduces the generation of alarm events by 60%. The use of 274 may or may not be proper from a process standpoint, but the effect should be understood.

9.6 After the D&R—Implementation of Changes

Changing the alarm system to one as specified by a D&R will usually involve a major shift in operating methodology for most operators. It can be quite uncomfortable for them, as well as staff engineers, to accept. There are several considerations and methods to accomplish effective implementation.

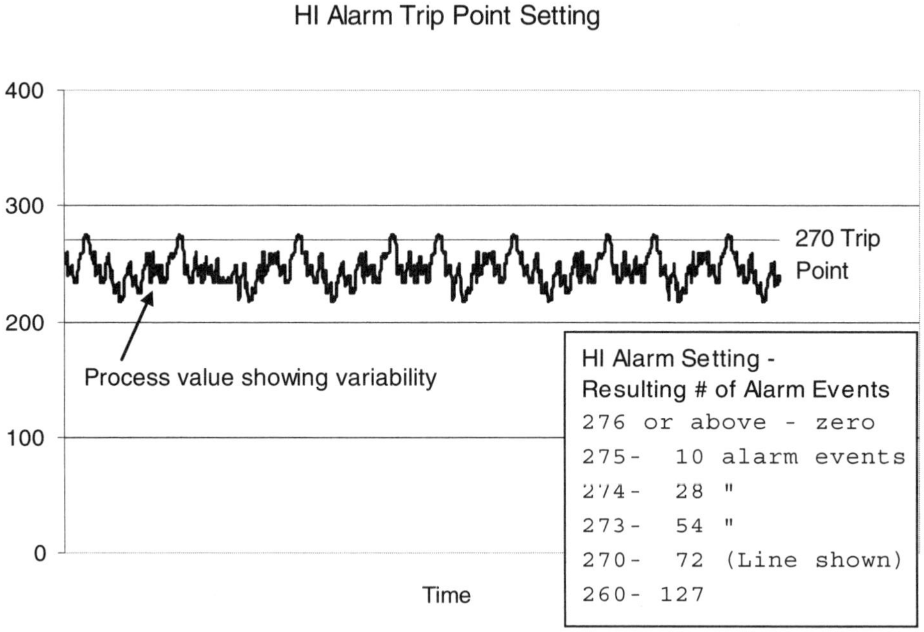

Figure 9–1 Process history and alarm set points.

Training

Implementation of D&R results, in the form of either a new installation or revision of an existing alarm configuration, requires operator and staff training. The training should cover several areas and include specific points:

- A review of the alarm baseline conditions that lead to the decision to perform a D&R
- A general overview of the alarm philosophy and assumptions
- Every alarm requires a response; it is unacceptable to ignore an alarm
- Every alarm is documented and has an identified response
- Alarm Priority is used to distinguish the order of response to alarms
- Rules and procedures regarding handling and reporting nuisance alarms
- Features of the DCS alarm presentation, annunciation and management
- Permissible and non-permissible changes to the alarm system by operations (in particular that Alarm trip points and priorities may not be changed without following the proper MOC process)
- The specifics of the site MOC process as it relates to alarms
- Proper use of all alarm handling strategies, such as Shelving, State-Based, Flood Suppression, etc.
- Instructions for accessing on-line Master Alarm database information
- Alarm system performance metrics
- Proper and improper alarm suppression
- Alarm setting audit and enforcement

A one-time handout should be prepared showing the changes in the alarm system from the D&R, particularly changes to Emergency and High priority alarms. Training should be accomplishable in 3 to 4 hours per group.

Implementing the DCS Changes Specified by the D&R

A D&R will usually produce several hundred or more desired changes in alarm configuration. Most DCSs have the ability to accomplish such changes in bulk, given a properly formatted file. Export of the desired configuration data from the D&R software is the logical starting point for this method. Proper site MOC procedures must be followed for implementing these changes.

It is possible to implement many changes with the use of the Alarm Audit and Enforcement software. Activate it with the new Master Alarm Database and the changes will occur in one cycle of the software.

Note that not every desired alarm change can be accomplished in such an on-line fashion. Sometimes, depending on the point type, the specific change, and the DCS, a point may have to be taken off-line to accomplish the change, and then reactivated. Care must be taken that the process is not disturbed during these operations.

Be sure to read "Point and Program References to Alarms" (Chapter 13, section 14) to avoid some significant and potentially major consequences in implementing D&R changes!

It is best to accomplish the changes all at once, by console, to the maximum extent possible. Do not spread them out over a period of days or weeks. Would you want your appendix removed one millimeter at a time over several days?

9.7 D&R—Expected Duration, Costs, and Helpful Advice

Assume that you have 2,000 alarmable points on a console, a typical, perhaps slightly low amount. Each point can generally have more than one alarm. We have identified that you will probably have 4 to 6 people in the room during the D&R, at a significant cost. If you are very experienced in performing a D&R, and are very well prepared, you can average about 100 points per day. Thus you will need 20 days of conference room time, adding up to about 100 person-days. There is preparation and post-D&R work to be done as well, outside of the meetings. During these 20 days, the operators are probably on overtime pay. The engineers are not, and the work is piling up back on their desks. Other improvements are being delayed. This is a non-trivial, expensive exercise!

Having said that, understand some assumptions and limitations about the 100 point-per-day number, because it can be quite optimistic.

- The rate is based on finding a substantial amount (at least half, on a point basis) of duplicate facilities in the system. This is where you rationalize 50 points on Furnace A, and then say "Furnace B is just like A—just copy the information over to the correct points." The copying is done outside of the meeting environment.
- If you do not have that many duplicates, you will likely not achieve 100 points per day. Why the need for so many duplicates? Take an 8 hour day, excluding lunch, and then subtract two, 15-minute breaks. That leaves 7.5 working hours, or 450 minutes. If you take 10 minutes per point to discuss all of the alarms on that point, you will rationalize 45 points per day. You can see the difficulty in having a 10 minute point discussion involving 5 people! You must stay focused and organized. Some points will take longer than 10 minutes.
- You will never accomplish the rate if you are trying to decipher complicated logic diagrams or listings. A person that knows the function of those systems must be in the room when they are being discussed.
- Make maximum use of the Specific Alarm Design Considerations section, to predefine settings for certain common alarm types, thus not needing to discuss them.
- The first day or two will probably have a low number of alarms rationalized as people become accustomed to the method.
- The more "nice-to-have" information you try to capture during a D&R, the longer it will take and the more it will cost. Be sure you intend to actually do something productive with the information, and not just capture it in another database that must be kept up to date.
- It is strongly recommended that you hire some highly-experienced help to facilitate D&R, at least for the first system that you do (or at least for the first few days), so you can learn the procedure and be most productive.

It has been our experience that you cannot get the D&R done in a shorter overall period of time by working longer hours each day. It is not the type of work that you can do for extremely long hours in a row.

Helpful Hints for Conducting a D&R

- Have food and lots of coffee in the meetings.
- Start on time and maintain on-time breaks. You MUST have breaks in doing this type of work.
- Do not allow pager and phone distractions.
- Have baseline information available about each point. During rationalization, it is very helpful to be able to say, "This next alarm is the 6th most-frequent one on the system, and frequently chatters." Of course, if you are following the advice in this book, you will have fixed such an alarm long before you get into D&R!
- Have large copies of the rationalization grids printed and hang them on the wall.
- Have a computer projector to display the D&R software as each point is discussed. It should also be capable of projecting process history data when analog point trip points are discussed.
- Try to capture only useful bullet items in the categories of causes and consequences. Much of what will be said, particularly early in the effort, is not very useful to collect. For example, if you are reviewing a high temperature alarm, the alarm cause of "Increase in temperature" is obvious and not useful. Useful information would be such things as
    - Loss of cooling pump
    - Fin-Fan off
    - Heat exchanger fouling
- The generic cause of "Instrument or Equipment Malfunction" is barely worth capturing, as the response to it is usually "Contact Maintenance."
- Try to stave off long rambling discussions or "war stories."
- Often the Corrective Action may simply be "Troubleshoot to determine exact problem." Consider a simple storage tank with three

inlets and three outlets. The complete and proper response to a high level alarm could run into several pages of alternative scenarios.

- Capturing a detailed troubleshooting procedure might be nice but you probably do not have time. (An example of really useful corrective action would be: "The most common problem causing the high tank level is the inlet stream from the condenser from weir overflow, not the bypass line from the cooler. Check the condenser and readjust its operation by altering the level controller and checking the source temperature...") If operating procedures contain such information, you will probably not want to duplicate it in the D&R database, because then there are two places that must be kept current and synchronized.
- Capturing the obvious is not useful. (Example: Corrective Action for a low flow alarm: "Increase the flow.")
- Capture action item notes in a separate document, not in the Master Alarm Database. These are items such as "This alarm should be put into a common alarm driven by logic with the other compressor vibration alarms." Or "The alarm on this point should be CHANGE-OF-STATE, not OFF-NORMAL."
- One person will also capture notes, usually by hand, to keep the D&R meeting moving. A frequent note will involve equipment duplications, such as "the D&R of point PC413 will be just like that of PC776." The computer manipulations to copy such information should be done outside of the meeting. Use a clear and consistent system for such notes; trying to figure them out 2 weeks later back in the office will be impossible if you do not.

Common Action Items from a D&R

The following items come up often during a D&R and result in action items.

- On a digital, change the alarm type from A to B or C where the choices of A, B, and C include OFF-NORMAL, COMMAND-DISAGREE, or CHANGE-OF-STATE. OFF-NORMAL is commonly overused and results in stale alarms.

- "Add Logic" to prevent stale alarms. This generally means building a logic point or program to improve the alarm and make it activate only as appropriate.
- Group into a Common Alarm (typically for multiple individual alarms, such as compressor vibration.)
- Build a proper diagnostic graphic that shows the status of all of the initiators of the common alarm point.
- Change Point Description to "-whatever-". This can really get out of hand during the D&R. In fact, meeting time should not be taken up with this; it can go on for hours. Our advice is that if this issue comes up, do a simple database query of point and description and paste the result into a spreadsheet. Give the spreadsheet to the people that want to change all of the point descriptions and say, "Off-line from the meeting you can go through the list and make all of the suggested point description changes that you need. Then submit the list to the control engineering group for implementation."
- Point needs re-ranging to avoid Bad Measurement alarm (or other types of point configuration changes, i.e., change deadband.)
- Change State Text of point, such as "RUN" to "RUNNING" and "STOP" to "STOPPED." The use of these terms in particular is often confusing. On a point where you are making the *command* to a machine, the terms "Start" and "Stop" are appropriate. On a point where the status of a machine is being shown, "Running" and "Stopped" are clearer.
- Needed Graphics changes

CHAPTER 10

Real-Time Alarm Handling Methods

"It is not only necessary to do the right thing, but to do it in the right way. And the only problem you have, is what the right thing to do and what is the right way to do it?"

—Dwight David Eisenhower

There are specific problems and best practices associated with alarm suppression, alarm shelving, state-based alarming, alarm flood suppression, auditing and enforcement of alarm settings, and operator alert systems.

10.1 Nuisance Alarms and the Dangers of Alarm Suppression

In a DCS, you intentionally configure an alarm on a point. You assign, on an analog point, a trip setting and a priority that causes it to annunciate to the operator. Most DCSs have another setting on the alarm that you can select—an alarm suppression setting. If suppression is "OFF" then the alarm works normally. If suppression is "ON", then the alarm behaves *as if you never configured it in the first place.* We call this situation "Alarm Suppression" (see Figure 10–1). This is a potentially very dangerous setting

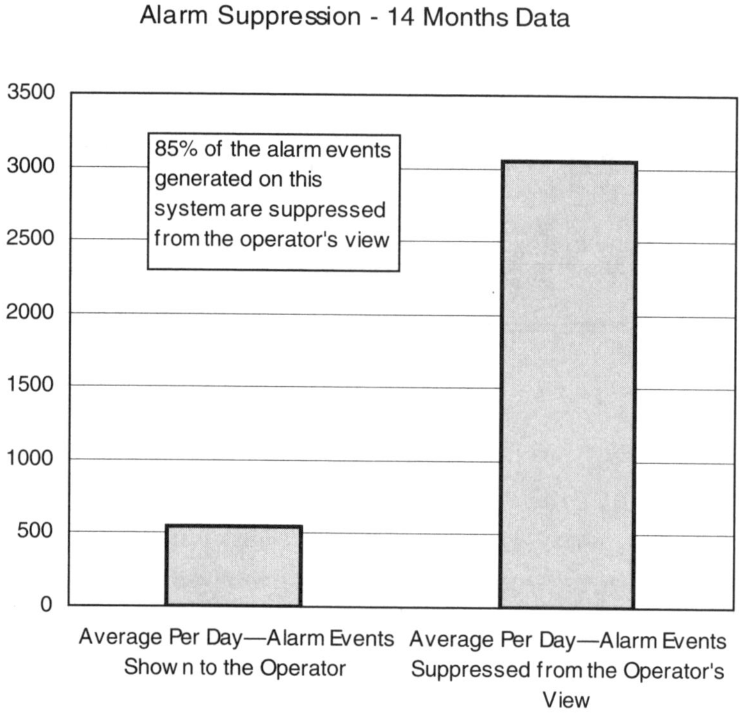

Figure 10–1 Alarm suppression in significant amounts.

to manage and we have very often seen it mismanaged in very hazardous ways. It is fundamentally different than "de-configuring" the alarm.

The intent of this setting is, naturally, benevolent. It was intended for use when a field sensor has a problem and begins creating a chattering alarm every few seconds. The operator needs and looks for a way to eliminate this nuisance. Otherwise they could endure continuous beeping and flashing and hitting of the Acknowledgement key for hours or days, and this will interfere with the proper response to other "real" alarms. Alarm Suppression is an "override." You could accomplish the same result by changing an alarm's priority to "No Alarm" or "No Action" or the equivalent provided on the DCS, but this method is made intentionally easier to use by the DCS manufacturer.

With a key, a password, or the correct access level, generally three simple keystrokes takes a fully configured alarm and *indefinitely* prevents it from being seen by the operator!

Some DCSs have good tools for managing the suppression setting. Ideally the access to it should be highly controllable, with a different access security level for this change than other changes on the system. It should take only a few keystrokes to call up the entire list of suppressed alarms and for each one, see when it was suppressed, for how long it has been in this condition, and the reason that it was suppressed in the first place. The DCS should periodically present the list for re-annunciation, and not let them be "forgotten about." However, such tools are not commonly standard. Instead, it is all too common that nuisance alarms are suppressed and then ignored. No written records are produced, no maintenance work orders for problem correction are submitted, no MOC procedures are followed, and no technical staff person is informed.

Many companies have procedures that say that, when alarm suppression is done, written entries should be made in various logbooks. But when you obtain and examine that list and what is actually on the DCS, you find that such written procedures simply do not work. If you have such procedures, you should go and pull the real suppressed alarm list straight from the DCS and see if it matches the written records. Prepare to be surprised. Then worried.

Studies of a number of major incidents have shown that suppressed alarms were a contributing factor to the operator's inability to properly assess the situation and take corrective action. Further investigation of these incidents reveals key alarms had been disabled and forgotten for an extended period of time. Imagine going before a jury and trying to explain why the alarm had been disabled at 2AM, 4 months before the incident, and no one since did anything about it. It could be costly.

Could this happen to your plant? Has it already happened? Has it been checked lately? What could happen if it's not dealt with?

Besides needing very tight control over alarm suppression, you should also have an iron-clad requirement that operators must obtain and check the list of suppressed alarms at the start of every shift. This should be a procedural requirement, and an easy-to-use, accurate method must be provided for their use. "Reviewing logbooks" will not work. Try it yourself if you have such procedures—check the written logbooks as if you are an operator returning from a 2-week vacation, and try to quickly and reliably determine the currently suppressed alarm list and repair status.

We can hear the grumbling now from the Production department. "Another new requirement from some know-it-all. We are doing just fine with our system now." Well, this advice is based upon many years of experience working in Production. There are lots of different pressures and conflicting priorities for production engineers. They can become really annoyed when someone without those responsibilities would come up with some new "good idea" for Production to have to do. Particularly when the idea came from staff groups, or even worse, outside "consultants" who had probably never made a pound of anything in their career! So we are really, really sympathetic with the Production point of view!

Now, having said that, if you are going to have an alarm, it is something upon which the operator relies—the operator expects it to be working. It is also something to help the plant manager sleep well at night—the alarm is there for a reason. If you take an alarm out of service, it is important that any operator that comes on shift knows that it is not there anymore. And it doesn't matter if the operator came in the day before or if they are just back from an extended vacation or sick leave—the operator must easily know (and must be *required* to know) which alarms are out of service. And the plant manager, to sleep well, must feel that there are methods and procedures in place that will always result in the operator's knowing an accurate list of the alarms not in service. Suppressed alarms must not be forgotten about, or left in that condition indefinitely; they must be fixed.

So to our friends in Production—you are free to not require this practice. But if you have an incident where in any way whatsoever a suppressed alarm played even a minor role, you are going to be held accountable, with lots of explaining to do. After an incident you will have a very difficult time explaining to upper management why you did not think it important to require that, when the operator comes on shift, they know which of their alarms are not in service.

It is because we are sympathetic with the pressure on Production that we put this forward strongly. If you have even a minor incident, you will probably wind up with complex externally-mandated procedures, audits, and so forth if this requirement is overlooked—procedures you will like much less than ones you could efficiently come up with yourselves based on this requirement.

The answer to this problem is proper, controlled, Alarm Shelving.

10.2 Alarm Shelving

Malfunctioning individual alarms, or groups of associated alarms may need to be suppressed for temporary periods. Such suppression must be controlled to ensure proper re-activation. Alarm Shelving is defined as the collection of techniques used to temporarily suppress alarms in a controlled manner, meeting several different and important criteria.

Manual (paper-based) shelving/unshelving procedures and processes can become too cumbersome if alarms are shelved too frequently or if there are too many shelved alarms to effectively track and unshelve. Paper-based systems have proven to be unreliable throughout industry. If manual procedures and/or processes are too cumbersome or unreliable, then automated alarm shelving solutions need to be implemented.

Regardless of whether manual MOC procedures or automated solutions are implemented for a particular area of the plant, alarm shelving must be done in a way that meets the following rigorous requirements.

It is essential that operators must know, each shift, which alarms have been removed from service and for how long. Shelving solutions must easily supply this information and operators must be required to query the system upon shift change and prior to startup of equipment. This includes obtaining the knowledge of points that are inactivated and points with malfunctioning sensors. Appropriate tools should be supplied to facilitate this process.

Alarm Shelving cannot be indefinite.

One of the major purposes of a shelving system is that it must not let the fact that alarms are shelved be "forgotten about." Periodic notification deals with this—like a snooze alarm. Your snooze alarm would not be acceptable if it had a 100-day setting, not a 10-minute setting. The shelving system should work like this—it should say (generally once a day or once a shift):

> *"Hey there operator—I have these 14 shelved alarms—either snooze them all for <whatever time you configure> or I will now bring them back online. Or, you can pick which ones individually to snooze or bring online."*

It is not acceptable that a shelving solution produce an alarm flood by automatically reactivating alarms without the operator's knowledge that it is about to occur.

Shelving solutions must be controllable by priority, with the ability to set time limits or require certain approvals. An example of a best practice in this regard is:

> Shelving of an Emergency or High Priority Alarm requires Supervisor approval and requires renewal at the start of every shift.

A shelving solution must work in proper coordination with other alarm handlers such as State-Based, Flood Suppression, and Alarm Settings Audit/Enforce mechanisms.

Manual and/or automated alarm shelving methods should be used in a consistent manner and should suppress only the specifically desired alarm functions. In other words, do not shelve a specific single alarm by a method that suppresses all of the other alarms on a point.

Chattering alarms should be prevented from being recorded to an event historian when shelved. If a chattering alarm is shelved but is still recorded in an event historian, it will continue to fill the alarm journal with up to thousands of alarm events over short time periods. This would flood the event recording system with alarm traffic, which may slow the speed of system response as well as saturating journals with alarm noise.

10.3 State-Based Alarming

Most alarms in a process unit pertain to the normal operating state of a piece of equipment. However, equipment often has several normal, but differing, operating states. DCS alarm capabilities are normally only for single-state, single-value trip points and priorities.

A few common state examples include:

- Running
- Not Running
- Startup
- Shutdown
- Full Rates
- Half Rates
- Both Trains Running
- Single Train Operation (i.e., Train "A" or "B" is shutdown)
- Fuel or Raw Material Type "A" or "B" or "C..."
- Making Product "X" or "Y" or "Z..."
- Coking or De-Coking
- Hot Standby (spinning reserve)

Besides individual pieces of equipment, entire sections of an operating unit may have different operating modes where fixed alarms produce inconsistent results. For example, the process may run in modes where certain sub-sections are intentionally shut down, producing a variety of alarms. Or, redundant equipment may produce alarms when unused, even though that is a "normal" and proper operating condition. In these circumstances, the alarms produced do not meet the real criteria for an alarm (there is no operator action to take) and will become stale and contribute to alarm floods and confusion.

It is a best practice that all such normal conditions occurring within various operating states should not cause alarms. Alarms should be produced only upon abnormal or unexpected events. State-Based alarm methodologies produce dynamic alarm configurations based upon the specific process and equipment conditions (see Figure 10–2). Changes to subsets of alarm trip points and priorities are configured for appropriate alarms and enabled based on detection of plant state.

While particularly important in Batch Processes, this is also appropriate for steady-state processes that incorporate different feedstocks, rates, or have different operating modes.

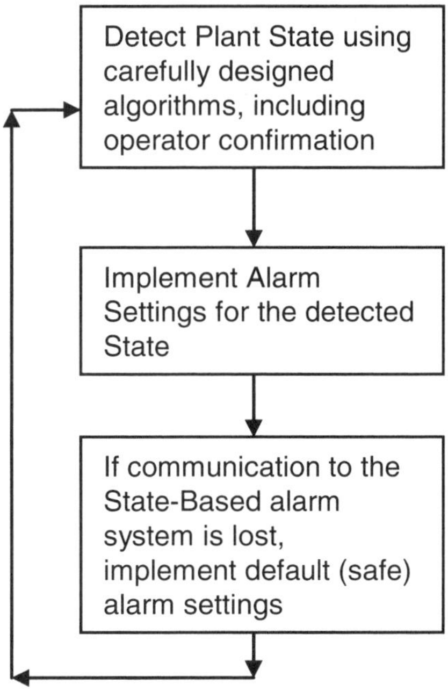

Figure 10–2 State-based alarming.

Two software components are required for handling State-Based alarms: a State Detector and a State Enforcer. The Detector uses available information (which should usually include operator confirmation) to correctly identify the current operating state of the equipment, while the Enforcer actually makes the desired alarm modifications. Neither, either or both of these tasks may be automated.

Guidelines for Implementation of State-Based Alarming

If multiple process states producing differing alarms are identified, these must be documented during alarm rationalization (D&R). State transitions requiring alarm system modifications should be handled by one of the following methods:

- Semi-Automated transition, utilizing the operator to identify or confirm the correct state and initiate the change
- Fully automated transition, with no input required from the operator
- Manual transition, with changes identified and performed individually by the operator

Documentation and other indication must be provided to communicate the current operating state to the operator. Any software methodology for dynamic change of alarms must be robust and have fail-safe mechanisms, in which alarms are restored to the default (steady-state operation) state if the Detector or Enforcer fails. This is typically accomplished through a watchdog timer setup. Obviously the operator must be informed that the state-based alarm system has failed. So (you guessed it), the situation is alarmed. We recommend a countdown-type alarm, such as "The State-Based Alarm System has stopped communicating. You have 15 minutes to get it working again before I reset everything to the default alarm setting values."

State Detection Logic

A key to proper state-based alarming is the configuration of the State Detection Logic (SDL). The purpose of the logic is to read and use appropriate process values on the system for determining the current state of the plant, train, vessel, or subsystem.

- State Detection Logic (SDL) should use values from at least two sensors, one primary and one confirming.

- The sensors chosen should be *very* reliable.
- Operator confirmation of the "proposed" detected state should usually be used to gain confidence, and then the "training wheels" should be removed.
- The SDL should incorporate proper handling if one of the sensors used goes to a Bad Value, Low Confidence, or otherwise becomes unavailable.
- The concept of deadband must be properly incorporated when using analog values.
- The SDL must not "chatter" and must be specifically designed to avoid chattering!

There are limitations to the speed at which SDLs can operate on most systems, related to the communication protocol used, available bandwidth, and the computational platform involved. A cycle of 10 seconds is practical and accomplishable. A fast cycle such as once per second may be difficult to achieve reliably.

Recommendations for "Shutdown State" Alarm Settings

In the case of equipment or entire process trains being intentionally shut down, the first thought might be that most or all of the alarms in the operating state should simply be de-configured or suppressed. This would be very, very wrong!

When you take all alarms out of service during a shutdown you implicitly assume proper isolation and no contamination of stored material. With all alarms out of service, isolation failures, unexpected reactions from contamination, etc. cannot be detected. There is at least one known incident of a chemical plant releasing materials and sending neighbors to the hospital, because the unit's alarm system was "disabled" when an isolation failure occurred while the unit was shutdown.

The proper alarm settings for out-of-service equipment have to do with immediate detection of any return to a higher energy state, or loss of equipment isolation. While low-energy alarms might be eliminated, the high-energy alarms should be retained, but set to activate at much lower thresholds. Figure 10–3 is an example.

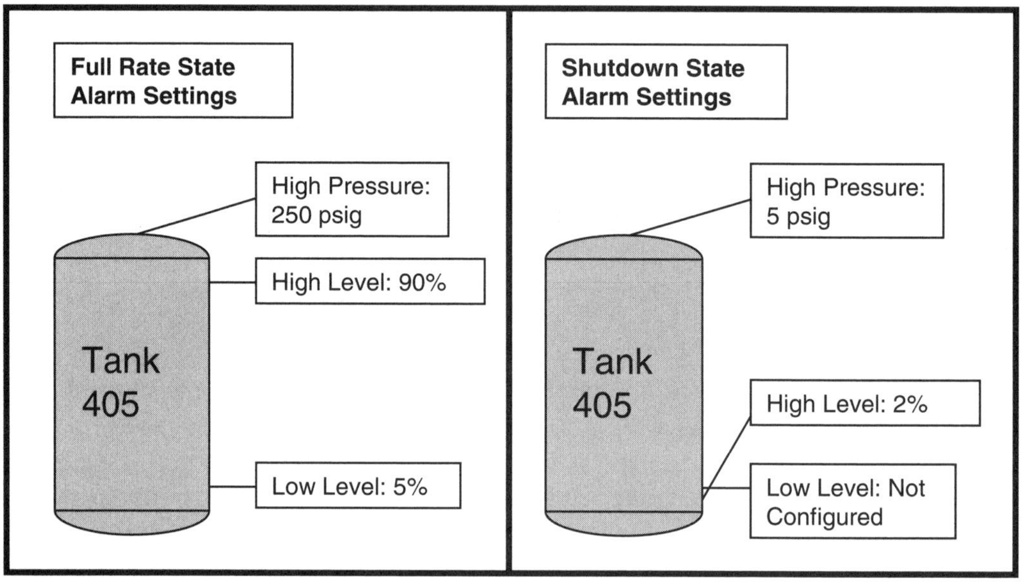

Figure 10–3 Proper alarm settings for the low energy state.

With the proper "shutdown state" alarm settings, should an isolation failure or contamination occur, the operator is informed and can take proper corrective action.

The same methodology for altering alarm system settings based on operating state can be extended to other aspects of process control. Parameters such as controller setpoints, modes, tuning coefficients, and so forth are candidates for dynamic change—but beyond the scope of this book!

10.4 Alarm Flood Suppression

Alarm floods can make a difficult process situation much worse. In a severe flood, the alarm system becomes a nuisance, a hindrance, or a distraction, rather than a useful tool. The risk of major process upset or accident is much higher during an alarm flood.

Flood Suppression is the dynamic management of pre-defined groups of alarms based on detection of equipment state and triggering events. The same engine and techniques for state-based alarming are used. The most common cause of alarm floods, once nuisance malfunctioning alarms have been eliminated, is the inadvertent shutdown of a piece of equipment—such as a compressor.

Design of a Flood Suppression Strategy

Consider a compressor. While running, it has many diagnostic alarms that are relevant. Low speed, low oil pressure, high oil temperature, and the like are all important alarms if the compressor is running. If such alarms are properly responded to, a compressor shutdown can be avoided.

However, if the compressor trips off for any reason, many of these alarms will activate under those circumstances as well. In such a case, they are not relevant to the operator at that time. The important action for the operator to take after the trip is to properly manage the upstream and downstream effects to minimize the resulting upset. It does the operator no good to get the "low oil pressure" alarm on the compressor when it has just tripped off because of high level in the suction knockout drum! There are usually several dozen such alarms that can be identified, that will always activate after an inadvertent trip. These are the alarms to be modified or suppressed in the flood suppression strategy. By eliminating these alarms, the distraction to the operator is minimized.

The use of alarm records from prior such trips, as well as P&ID examination, will supply the candidate list for a flood suppression strategy for a particular piece of equipment.

The state detection logic (SDL) to be used must be carefully thought out. The logic should detect the transition that "The compressor *has been running* but *has now stopped*." The logic is not just "The compressor has stopped." The logic result must not chatter! A flood suppression SDL may be one where you do not want to include operator confirmation of the event, so that it can activate as soon as possible to minimize the flood. In such a case, the SDL must be very reliable.

10.5 Auditing and Enforcement of Alarm Settings

Alarm Audit and Enforcement software is essential. It is used to periodically and automatically check for changes from the proper alarm settings (as determined from D&R and contained in the Master Alarm Database). The software should report such changes, and restore the system to the proper settings. DCS alarm settings are so easy to change that without such mechanisms it is virtually guaranteed that the system will quickly drift into a suboptimal state. We have seen the configured alarms in a system increase by more than 50% one year after a D&R, due to poor management-of-change control.

Paper-based MOC systems will consistently fail in keeping your alarm settings proper. Enforcement software ensures compliance with your MOC system. It is critical that the enforcement software communicate with and understand all of the functionalities of any state-based alarming, flood suppression, or alarm shelving strategies in use. Obviously, enforcement should be accomplished with settings based on the current state of the process and should not reactivate alarms that are properly shelved. If the enforcement software does not work in this way, you will have chaos.

10.6 Operator Alert Systems

Now put yourself in the position of the operator:

> *The alarm system has been "locked down"—I can no longer adjust any of the settings. Today, I need to transfer material from the field tank to the unit tank. The transfer will take several hours, and I have to leave enough in the field tank to load 5 tank cars tomorrow.*
>
> *How do I know when to call the field pumper and stop the transfer? Do I have to check the level every 15 minutes? What if I forget? Those stupid engineers don't trust me to do the job that they pay me for! They take away the system I've used for years and give me nothing to replace it.*
>
> *I know—I'll set a timer on my cell phone to remind me. Darn, the battery is low—I hope it lasts. I would sure hate to miss those shipments tomorrow....*

Individual operators have a need for "on-the-fly" configuration of various system reminders and functions based on the process attaining certain values. Historically, operator change of the alarm system trip points has been allowed to be used for this purpose (and other similar purposes). However, the setting of individual preferences as alarm limits results in sub-optimization of the process, causes shift-based process variation, introduces non-rationalized alarms, contributes to alarm floods, and is therefore not in keeping with best practices.

Operator change of the alarm system limits is thus a practice to be avoided. In fact, it should be actively prevented by automated Audit and Enforcement mechanisms.

Therefore, the operator needs a configurable set of tools with which to meet such normal operating needs. This tool set is called an Operator Alert System, which is totally separate from the alarm system. Alerts fall outside of the alarm priority-setting methodology and must have their own audible tone. Alerts must not appear on the Alarm Summary screen.

Alerts have the following characteristics:

- Alerts are user-configurable (operator or shift) and user-controllable.
- Alerts allow operators to monitor important process variables for steady state control or for other purposes.
- Alerts, unlike alarms, can be safely set aside during abnormal or upset conditions when the response to alarms must predominate.

It is not wise to use low-priority alarms as operator alerts. In such cases, the low priority alarms become a mixture of rationalized, known-action-to-take real alarms, and unrationalized, unknown-action-to-take annunciations that sound exactly the same. The alarm system is *intentionally* designed to be an interruption to the operator, with known tones and behaviors. ("We interrupt our normally scheduled chemical process to bring you this important announcement!") Therefore the initial indication to the operator should not be the same "thing" for an alert as for an alarm. Would you want your telephone, pager, cell phone, alarm clock, doorbell, and microwave oven to share the exact same sound?

Provision of an Operator Alert System is a best practice. If the alarm system is "locked down" per best practices, and an Alert system is not provided, then the operators will come up with a workaround. It is unsettling to see a row of cheap kitchen timers used as an integral part of a several-million dollar control system!

10.7 Alarm Handling Software—Write Your Own?

Companies sometimes consider writing their own DCS software to accomplish various alarm handling tasks. This is particularly true if there is a very talented in-house engineer or programmer interested in working on the problem. We have witnessed many examples of home-grown or "one-off" applications that eventually failed. In failing, these efforts resulted in costing more money than if proper commercial software had been obtained in the first place.

In general, the talented people that created such solutions, with the absolute best of intentions, were able to come up with a solution that worked—for awhile. But the press of other important job activities generally resulted in software that had these characteristics:

1. Poor error handling. Software must gracefully handle, and recover from, errors in user input or activities (users will do some crazy things), unexpected process values, values going to an unexpected state or range, temporary loss of connectivity, system load bursts, storage space limitations or "unexpectedly full" conditions, points being rendered inactive or modified, and a host of other problems. It would be very unusual for one internal-to-the-company person to have knowledge of all of the possibilities for errors that must be properly handled. On the other hand, an experienced commercial software supplier should have experience on hundreds of systems and configurations, and will have been through teething pains in encountering many unusual field conditions.

2. **Poor error reporting.** When something goes wrong, it can be very difficult and time-consuming to figure out "what." In-house developed software may not include creation of error or status logs and detailed diagnostics or breakpoints.

3. Poor longevity. After the original author is either transferred to another department or project, their time available to continue to support the software will drop to near zero. The first unexpected problem, or slight modification needed, will kill the entire application. The track record of home-grown DCS programming surviving the departure of the author is very, very poor.

4. **Poor adaptability.** Version changes in the DCS platform or hardware or system software can cause the need for modifications or code rewrite. Often the resource is unavailable to do this.

5. **Poor documentation.** While instructions for starting, stopping, and using a program might be adequate, the explanations of the working of the source code are often very inadequate. The ability of anyone else to modify it for a new or slightly altered situation could result in essentially having to start from scratch again.

6. Poor version control. Lack of dates, version numbers, explanations, and code controls are common.

7. **Poor User Interface.** A polished and consistent user interface can take as much or more time and effort to originate as the rest of the code and requires expert and specific programming knowledge and skills. It is unlikely that any UI created as a one-off situation will be very well-thought out from the user's perspective.

8. **Poor "growth" support.** After awhile, the ability of the in-house originator to remain interested in bug fixes or code adaptations will diminish.

CHAPTER 11

The Death of the Lightbox

"Put out the light."

—Theodore Roosevelt, last words, January 6, 1919

Many people think that engineers are not emotional. They are wrong. It's just that engineers get emotional about some strange things that most non-engineers don't think about—like "lightbox" alarm panels!

The lightbox used to be the very definition of an alarm (See Figure 11–1). It was the sole method to show an alarm; therefore, if something was "on the lightbox," it was an alarm.

As DCS systems became available, and old pneumatic or early electronic control systems were converted to a DCS, the question arose—"What do we do with the lightbox?" All possible options were used—from keeping it as is to eliminating it entirely. In between those extremes are many options for parallel wiring of the sensors that input to the lightbox into the DCS as well. But the time has now come to evaluate the issue of the lightbox dispassionately, which will be difficult for some readers of this book! We will lay out the case for and against lightboxes. In a nutshell, we

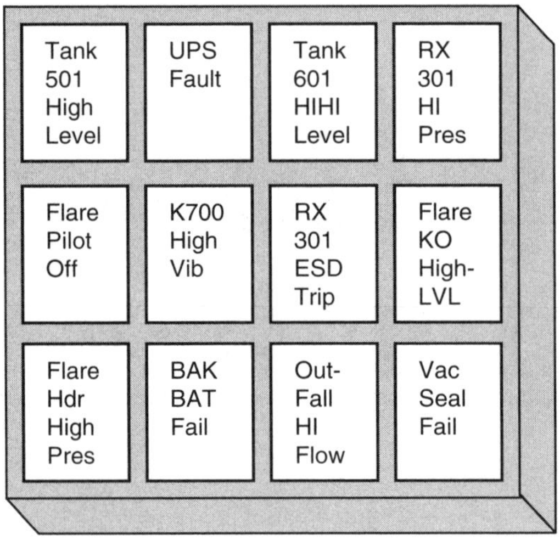

Figure 11–1 A typical lightbox.

think their time has passed. This section is placed after Alarm Rationalization so that you have a firm understanding of how an alarm's priority is determined. And certainly there is no higher priority alarm than one we choose to implement, quite expensively, on an external annunciator!

11.1 Advantages of a Lightbox

The lightbox is basically a very reliable light bulb for each alarm. That's it. The advantage is that a lightbox is visible at all times to the operator. It doesn't require the operator to call up a specific alarm display, and if all of the DCS screens "go dark" simultaneously, the lightbox is still there. It requires substantial effort to "defeat" or "disable" an alarm on a lightbox.

11.2 Observations

The primary reason for the lightbox has to do with a) engineering inertia, and b) its perceived reliability. This is contrasted to a perceived unreliability of a DCS in two areas—loss of view, and alarm configuration security.

11.3 Loss of View

The DCS "loss of view" phenomenon was quite real and unacceptably high in the early days of DCS development, up through the early 1990s. DCS reliability has seen major improvement. The electronics involved in getting the process information to the operator now have high redundancy and availability. Most operator consoles (i.e., span of control of a single human operator) have 4 to 6 screens, not only for practical use but also redundancy. If one fails, they limp along with 3 to 5 until it is fixed. It is common to have a single screen hardware failure about once a year, and replacement should rarely take more than a couple of days.

Total loss of view incidents are much less common than they used to be. But even if all the operator screens go dark, the control system continues to operate on the last setpoints and the safety system continues to function. It is a totally erroneous belief that a separate lightbox to show some alarms is a somehow adequate answer to a concern about loss of view. While common mode failure is real, a lightbox is not the answer. They represent a false sense of security; they do not make a control system measurably safer.

And, from a common sense point of view, if you have a total loss of view, it is difficult for the operator to effectively respond to an alarm on the lightbox. The reality is that the Safety Shutdown system will react as needed if the DCS screens fail and the process reaches shutdown limits. The answer to a problem with DCS reliability is *not* to add a lightbox! If your car engine sputters and misses, you don't think "I don't need to worry, I bought 3 spare tires!"

Incidentally, and anecdotally, we have reports that "loss of view" incidents have become very rare since the 1990's, and the cases almost always come about from someone working on or around the "Uninterruptible" power source (UPS) and not doing the work correctly. The very few incidents remaining seem to come from a catastrophic loss of power to the entire plant. In a personal experience of one of those, the nominally rated 30-minute DCS backup batteries lasted for 4 hours. Of course, with no power to any pumps, blowers, agitators, process heaters, and so forth, there wasn't a lot left to control, so shutdowns were initiated almost immediately after the power loss.

Perhaps the most worrisome current potential causes for loss of view are computer viruses, worms, spyware, and the like. PC technology has taken over the DCS, bringing with it low cost, easy adaptability, and all of the risks associated with complex networked computers. Rigorous control on operating system modifications, program installation, anti-virus methods, and firewalls must be in place and kept up-to-date. This is another hot topic in modern control, and beyond the scope of this book. But once again, the answer to worries about DCS security from viruses is not a lightbox!

11.4 Alarm Configuration Security

The ease of change of alarm configuration on a DCS is a significant concern. With the right access level and a few keystrokes, any DCS alarm can be suppressed or changed in a number of ways. And don't kid yourself that access keys, passwords, and "back doors" are not well known among those who work with the DCS! Section 11.5, "Auditing and Enforcement of Alarm Settings," addresses this issue, and such an enforcement mechanism is a requirement for an effective alarm system.

Disadvantages of a Lightbox:

- Alarms on a lightbox are expensive to implement.
- They are difficult and expensive to modify when re-arrangement (i.e., control relocation or consolidation) is needed.
- They have very limited functionality.
- The life cycle cost of a lightbox is very high.
- Long, expensive wiring runs are needed if they are located in a control room remote from the process.
- The lightbox provides no navigation abilities or enhancements to aid the operator in responding to an alarm.
- The lightbox provides no history of actuation, or for sequence-of-events, alarm causes, or operator responses, and does not easily integrate well with a DCS.
- And particularly, it is a best practice that the operator should have a single interface for interacting with alarms, and a single place and method for acknowledging alarms. Use of a lightbox will not achieve this best practice.

11.5 Proper Implementation of a Lightbox with a DCS

First, we will discuss the proper implementation of a lightbox with a DCS, *if you choose to have one*. Because having one is a choice; it is not a best practice requirement, as you will see.

A lightbox is limited to a very few alarms. Therefore the criteria for placing an alarm on the lightbox must be consistent, specific, and highly selective. We have seen hundreds of lightboxes and have made it a point to examine them closely. When the alarms are listed, we generally find little rhyme or reason as to how those particular alarms were chosen to be placed on the lightbox. We find mixtures of all sorts of different alarms with little coherency and consistency. Usually there were no consistent and logical guidelines; the selection was just based on the opinion of a few people, sometimes as a project afterthought. (If you have lightboxes, go and do these checks yourself.) The following are guidelines for effective lightbox implementation.

- Every alarm placed on the lightbox, must at least meet the rationalization criteria for the highest priority DCS alarm (i.e., "Emergency" in this book).
- Besides the rationalization test, there must be a specific reason and rationale by which the total group of configured Emergency alarms can be examined, and a proper subset of them determined for elevation to the lightbox. Note that *not* every alarm that comes out of a Process Hazard, or Safety Interlock Level, or Layer of Protection Analysis will meet these criteria!
- The lightbox alarms should be placed in a consistent, logical spatial pattern, and be well-labeled. The positioning, color, and other characteristics of the external alarm should aid the operator in performing the appropriate mitigating actions to the event.
- Lightbox alarms should be integrated with the DCS to allow a degree of history collection when activated.
- If a DCS alarm duplicates a lightbox alarm, then the DCS alarm should not be annunciated (and thus require acknowledgement) as well. But the process graphic should show the alarm condition properly.

11.6 The Death of the Lightbox

Our wide-ranging experience is that lightboxes are disappearing. New facilities and control rooms are designed and built without them. Control systems upgrade projects get rid of them. In pictures of new and planned control rooms, the lightbox is nowhere to be seen. The modern operator station includes large projection displays and looks similar to the bridge of a movie starship. You didn't see a lightbox on the *Enterprise*! The reasons for the demise of the lightbox have to do with all of the limitations that have been described.

Yet, if you mention this phenomenon as an issue to a group of control engineers, you will generate a lot of controversy! Lightboxes, as a tried and true technology, have their strong adherents. And in fact, you may encounter engineers that absolutely insist on them! The need is expressed from generally three points of view, each of which can be effectively countered.

> *The Instrument That Will Not Die*
>
> Maybe there's no lightbox, but if you look carefully, you will see an old pneumatic circular pen-chart recorder on the back wall of the Enterprise's *Galileo* shuttle. It seems those things will last forever.

11.7 Lightbox Rationales

Simplicity

You may hear it expressed that lightbox alarm systems are very simple, and that the various switches are easy to test. A simple high pressure switch as a source to the lightbox is often used as an example. This example fails the reality test. If a high pressure reading is critical enough that it deserves the lightbox alarm, then why would such an unreliable device as a pressure switch be chosen as the sensor? Certainly a pressure switch has more covert failure modes than a modern transmitter, is more difficult to calibrate, more difficult to adjust deadband, is more prone to cause a chattering alarm, lacks self-diagnostics—in general it is a poor choice for a critical sensor.

So if instead you use a modern transmitter (or redundant ones, probably, for such an important alarm) then you have to convert the analog signal(s) and somehow include the comparison to a trip point to make the annunciator lightbulb shine. Not quite so simple anymore—to install, test, document, alter, or troubleshoot.

Functionality in the marketing of technology always seems to trump simplicity. This is not an endorsement; it is a statement of fact about the environment we must manage. If simplicity were king, then cars would still have points in the distributor and that "electronic ignition" thing would have never caught on. Remember when Microsoft Word fit on one 320k floppy? (What's that from the back? "What's a floppy?" Ah, youth...)

Continuous Presence

The second point of view has to do with a desire to keep a certain set of alarms in view of the operator at all times. If you have this need, the better solution involves thinking outside the "lightbox" for a moment.

Simply design a graphic display that *looks* like a lightbox if you absolutely insist! Call it your "Critical Alarm Display." You can put just your most critical alarms on it, and you can make them have whatever behavior you want. And indeed you can give them *better* behavior than is possible with a lightbox. You can give them navigation abilities so that selecting one of them calls up (on a different screen) a more detailed display of the exact problem. You can include live updating process measurements and even trends. You can do all sorts of innovative things. You can (with most DCSs) assign this "Critical Alarm Display" to a button so that it is never more than one keypress away.

Or, if you must, you can *dedicate* part or all of a DCS screen to just that display and keep it displayed all of the time! Even at the prices of DCS displays, it is still much cheaper doing this than installing a lightbox. And as time passes and you need to add or modify critical alarms, it is much easier to do on the Critical Alarm Display than on a hardwired lightbox. This display method has much better functionality and lower cost than a lightbox. That's an easy choice.

Probabilistic Risk Assessment

Statistics: The only science that enables different experts using the same figures to draw different conclusions.

—Evan Esar (1899–1995)

The third point of view against eliminating the lightbox comes from the risk assessment realm. This book is not a treatise on proper calculation and design of safety systems; there are plenty of other books and standards on those. And there is a high amount of political correctness involved in discussing this area! But regarding the implementation of alarms, here is exactly what you need to know if an engineer comes to you and says that their calculations indicate that a particular alarm must be implemented on a lightbox rather than in the DCS.

There are many experienced and knowledgeable practitioners of these risk assessment and design methodologies. The methods are based upon probability and statistics. The assessment results are highly dependent upon the assumptions that are made at the outset and also throughout the assessment, at many steps along the way. Those assumptions have both objective and subjective elements. Equally knowledgeable, respected, and experienced engineering practitioners can evaluate the same system and come to different conclusions; in fact, this is not only possible but likely. Many credible engineers have concluded that reliability and independence requirements for most hazard mitigation scenarios can be achieved either with a lightbox or with an equivalent DCS interface.

And indeed, because there is such wide variation in the valid opinions of knowledgeable people, the relevant standards were intentionally written with lots of room for subjective interpretation. The reality is that you can use these methods, if you must, either to justify adding (or keeping) a lightbox or to eliminate (or resist adding) a lightbox. This is, perhaps, an uncomfortable state of affairs for logical-minded engineers, but it is a reality. Having a lightbox is a *choice,* not a requirement.

And the proof is clear—refineries and chemical companies handling highly flammable and hazardous materials are implementing new facilities and control systems. They are using the latest advanced engineering calculation methodologies and methods (and spending plenty of money in doing so). If there was anywhere that you would think that lightboxes would predominate, if they were actually needed from these probabilistic risk assessments, it would be here! Yet the facilities are being built and operated without hanging boxes of lightbulbs on the wall. The truth is they just aren't needed.

There is a lot to be said in favor of the old, big, spread-out, see-everything-at-a-glance panelboard controls and lightboxes. They are similar to the "steam gauges" in airplanes that many pilots are familiar with. But all the new planes now being built—even small propeller ones—have full, software-driven glass cockpits. The panelboards are not coming back.

11.8 Conclusion

The DCS should be used to present alarms in a consistent fashion. The use of an external, segregated alarm display is not an industry best practice. Historic reasons for the inclusion of lightboxes are generally no longer valid. They are expensive, have very limited functionality, and do not integrate well into a modern DCS.

CHAPTER 12

Specific Alarm Design Considerations

"You must learn from the mistakes of other. You can't possibly live long enough to make them all yourself."

—Sam Levenson

There is no need to individually discuss many types of alarms, on a point-by-point basis, during D&R. Instead, prioritization rules for several types can be initially agreed on, then used. Here are several categories of such alarms, and the best practice configuration and priority of those alarms.

12.1 Flammable and Toxic Gas Detectors

In the D&R section, we stated that you will likely have very few alarms where the failure of an operator to respond to the alarm could result in a person getting hurt. This section covers the most common of those few instances. In the case of Flammable and Toxic Gas Detectors, the operator gets the alarm and the primary response is to take actions that warn and ensure that anyone in the affected area leaves immediately. Time is of the essence. If the operator does not make this response, then people in the area could be injured. Therefore, Flammable and Toxic Gas Detectors shall all have Emergency Priority alarms.

Such alarms should be displayed on a graphic that indicates their geographic location. Most companies add wind direction and velocity indicators in the same graphic. In some cases, depending on the hardware, malfunction alarms on the detection hardware should have High (rather than the default Low) priority alarms.

12.2 Safety Shower and Eyebath Actuation Alarms

Many companies alarm the actuation of safety showers and eyebaths. The rationale is that someone has been seriously exposed and needs assistance. Failure of the operator to respond and ensure assistance is immediately sent could result in further harm to the person. Therefore such alarms should also be Emergency priority.

12.3 Building-Related Alarms

In some cases, alarms of the following types (particularly for outlying or unoccupied buildings) may be wired into the DCS and handled by a central control room operator. Recommendations for priority are as follows:

Emergency Priority (for the same personnel safety reasons)

- Smoke and/or Fire
- Carbon Monoxide
- Presence of Hydrocarbons (e.g., Lower Explosive Limit—LEL)
- Low Percent Oxygen (insufficient breathable air—sometimes used in buildings that contain pressurized inert gas sources)
- Health or Status malfunction alarms regarding the detection hardware in this category should have High priority alarms.

High Priority

- Failure of an important device's redundant power supply
- Uninterruptible Power Source (UPS) malfunction

Low Priority

- Building or Cabinet Pressurization (consider on-delays to prevent nuisance alarms when opening building or cabinet doors)
- Temperature/Humidity (such as would indicate loss of HVAC in a room with computer equipment)
- Health or Status information regarding the detection hardware in this category shall have Low priority alarms.

12.4 Alarms from Instrument Malfunctions

It is quite common to see instrument malfunction alarms (such as Bad Value and others) appear in the Top 10 list of most frequent alarms on a system. When considered objectively, this is quite puzzling. Was the instrument *designed* to spend much of its time with a Bad Value? Of course not. So why are they there? Why are there typically from hundreds to thousands of alarm events weekly related to Bad Value? If you had asked your very best control engineer to intentionally design an indicator that would produce hundreds of bad value indications weekly, could they have done it? Doubtful, yet this situation is common throughout industry. For a thorough discussion of this problem and solutions, see chapter 13, "Common Alarm Problems and How to Solve Them."

Since you paid good money to install an instrument that is needed to control the process, it is reasonable to configure an alarm to notify the operator that the instrument is no longer doing its job. The action to be taken is normally a limited amount of operator troubleshooting. If the troubleshooting is unsuccessful, then the decision must be made as to whether maintenance callout should be obtained, or if the more routine writing of a work order is appropriate; the decision depends upon the criticality of the specific instrument.

Here is a set of best practice rules for instrument malfunction alarms.

- By default, all sensor points will have Bad Value alarms.
- Bad Value alarms are all Low priority (being consistent with a likely response time exceeding 30 minutes.) We recommend a minor modification to the rule, wherein the Bad Value alarm is configured to High priority if it is on a sensor that feeds an Emergency priority alarm. These sensors would be those for which the action involves immediate maintenance callout regardless of the time of day.

- In the event that the DCS supports 4 levels of fully annunciated priorities (different assignable colors and sounds), then a good practice can also be to isolate many of the Bad Value and similar diagnostic alarms into a 4th (lowest) priority used only for them. The 4 priorities become "Diagnostic, Low, High, and Emergency." This 4th priority would be used for sensors whose malfunction response is to simply write a maintenance work order for a normally scheduled, non-immediate response. If used, this ensures that all alarms in the Low priority remain those for which operator response within 30 minutes is needed, and "write-a-work-order" diagnostics are not mixed in with them. This separation helps the operator in high alarm rate situations, as the "Diagnostic" priority can be safely ignored for awhile.
- Where appropriate, group related Bad Value alarms into a common alarm, then provide a detailed display that shows the particular sensor involved.
- If an indicator point is an input to a controller point, alarm the Bad Value on the controller point, not the indicator point (since the controller is where operator action will be taken in response to the alarm).
- Bad Value alarms often propagate through several points. These must be revised so that a Bad Value event produces only one, not multiple, Bad Value alarms. There are generally several techniques to accomplish this based on the point type.

12.5 Use and Abuse of Combination Alarms

Combination Alarms (sometimes called Pre-Alarms) are where Process Value (PV) HIGH or PV LOW alarms are configured with, and often followed immediately by, another alarm (PV HIGH-HIGH or PV LOW-LOW). Some systems even have HHH and LLL capability. Such pre-alarms can contribute significantly to alarm flooding. Remember, just because something is possible to do in a DCS does not mean it is a good idea.

Systems are often initially configured "by default" with many or all pre-alarms configured. Often rules of thumb, rather than the principles of alarm rationalization, are used for the settings—rules of thumb such as to set all analog points with alarms at 10% for LL, 20% for Low, 80% for High, and 90% for HH. This is a very poor practice.

Some engineers always configure all such possible combinations, thinking, "If the operator misses the PV HIGH, which he might because he gets so many alarms, he still has a chance to see the HH (or even HHH) before something bad happens." This is like shooting yourself in the foot to distract yourself from the pain of your broken arm. It is why our favorite term for these is "Nag Alarms."

Examination of data will usually show you that many points with such alarms always have the HIGH and HH (or LOW and LL) alarms go off closely together. This indicates either:

- the process varies too quickly for the operator action taken at the HIGH alarm to prevent the excursion from reaching the HH alarm trip point, or
- the operator does not take action at the HIGH alarm but waits for the HH, or
- the trip point settings are too close together, or
- some combination of the above.

The proper, best-practice answer to the situation is often seen as drastic. It is to use combination alarms only under the following conditions:

- The operator actions for the first alarm (HIGH) vs. the next alarm (HH) must be *significantly different in kind or in degree*. In other words, do not alarm twice for the operator to take the very same action.
- There must be enough time after the pre-alarm to perform effective corrective action before the process activates the next alarm.

The application of these principles during D&R usually results in 90+% of these combinations being eliminated, as they fail the first principle above. The engineer and operator reaction to that is usually apprehension, because in an overloaded alarm system, these combinations do give the operator "one more chance" to see the situation and pick it out of the noise.

But you have to break that old paradigm, which is implicitly based on the assumption that the alarm system can never be brought under control, it will always produce too many alarms for the operator to really deal with individually, and the operator will always be forced to ignore some alarms due to their sheer volume. This is not true! You must be firm in resolving that one of the original assumptions—that operators will respond to all alarms—will be followed. It must be unacceptable to ignore an alarm, and the assumption that alarms will be ignored is *not* justification for adding more alarms. That just makes your problem worse, not better.

12.6 Redundant Sensors, Voting, and Shutdown Systems

Proper application of safety and reliability standards may well result in the installation of double or triple redundant sensors and alarming in some instances. This may or may not involve "voting" systems. A lot of attention must be given to the proper configuration of alarms from such systems. During process upsets and abnormal situations, improper multiple alarming may produce alarm floods. In addition, routine shutdowns may result in the activation of several standing alarms. Such nuisance alarms must be avoided if at all possible.

Consider that all of the following sets of alarms might be configured, by those unknowledgeable in Alarm Management, on a simple dual-sensor, one-out-of-two vote-to-trip interlock. Many of these alarms may be generated in the Safety System and then mirrored in the DCS via a communications link.

Interlock Voting Alarms

- Sensor A Pre-Trip (Safety System)
- Sensor A Pre-Trip (DCS)
- Sensor B Pre-Trip Alarm (Safety System)
- Sensor B Pre-Trip Alarm (DCS)
- Sensor A vote to trip alarm ("I have cast my vote to trip!")
- Sensor B vote to trip alarm ("I have cast my vote to trip!")
- Alarm for: "The Votes Have Been Counted, the Trip has been Voted Successfully!" (We have seen this, the "Precinct Chairman" alarm, configured even on a 1-out-of-2 interlock, *besides* the vote to trip alarm!)
- Alarm for: "Compressor Shutdown Actuator is taking action to shutdown the compressor" (This alarm occurs a few milliseconds after the Precinct Chairman alarm)
- Alarm for: "Compressor has now tripped!" (This is the overall trip Notification Alarm.)

Iinterlock Diagnostics

- Sensor A and Sensor B are differing in their readings by more than X%
- Sensor A out-of-range condition
- Sensor B out-of-range condition

Interlock Restarts

- The problem that caused Sensor A to vote to trip has gone away and thus you may reset this shutdown logic in order to restart. (This is the "I am ready to be reset" alarm.)
- Same as previous but for Sensor B
- The Reset for this shutdown logic has been activated. (This is the "I have been reset" alarm.)

The situation gets worse with three sensors and more complicated interlock actions.

Best Practice: All redundant and voting installations must be designed and reviewed on a case-by-case basis to ensure:

- Minimal multiple alarms result from process deviations.
- The operator will not receive a flood of unnecessary alarms during routine startup, shutdown, or other periods when the hazard scenario is not valid.

The case-by-case review of these redundant installations may require further study outside of the normal alarm system D&R. Safety considerations inherent in these redundant installations may necessitate logic-based alarming to be incorporated in the logic solver equipment, and not in the DCS.

If there are duplicate or similar analog measurements input to both the DCS and the Safety System (for control and trip determination, respectively), do not provide multiple alarms from both sources for the same process condition. The reading not alarmed may be displayed on the graphic. If the Safety System reading is chosen to be alarmed as a pre-trip indicator, ensure that its associated display ability goes to the correct DCS graphic for taking pre-trip corrective action.

An analog measurement selector (high, low or middle of three) equipped with a single pre-alarm is often a better design choice than multiple and separate A, B or C signal voting alarms. An alarm on a selector will generate the single, definitive, and earliest warning of an impending problem.

Valve position switches on ESD shutoff valves are often set to alarm when the valve performs the ESD action. This is incorrect; there should be an alarm only when the valve did *not* perform the proper action, so that the operator can take further action to ensure that the post-trip isolation is correct.

Diagnostic measurements on deviations between multiple sensors in a safety system can greatly increase the system reliability that comes from increased diagnostic coverage. This can in turn help meet the safety system reliability requirements with less capital investment and lower maintenance—even extending the time between required system maintenance turnarounds. However, it is not necessary to alarm each brief mismatch that occurs, only those that last for continued periods of time.

Bad Measurement alarms on devices connected only into the Safety System should not produce a general Safety System health/status alarm when activated. Such alarms should annunciate in the DCS as a Bad Measurement alarm on a point representing the point in the Safety System.

Good diagnostic graphics showing the state of the sensors and the voting process are the key for the operator to understand where the system is relative to a shutdown. Otherwise you are expecting the operator to keep "in their head" all of the details about the complex logic—for example, if a bad value is or is not a vote to trip of a particular sub-system. This is highly problematic. Can you always remember all the internal logic of how your automatic car door locks work? ("Let's see... if I click the button and close the door with the key still in the ignition, but the motor *is* running, will I be locked out? Heck, I'll just leave the door cracked open when I get the mail." Reading the manual is hopeless!)

12.7 Safety System (ESD) Alarm Priority

"The real crisis is not the crisis.
The real crisis is what you do before and after it."

—David Gerrold

Many engineers think that every Trip Notification Alarm ("The ≤*whatever*≥ has been shut down!") should be Emergency Priority, especially if it is generated from the Safety System. This is most assuredly not the case!

It may well be that proper alarm D&R shows that the pre-alarm to a trip might be a High or Emergency priority alarm, but the trip notification alarm itself (i.e., "The shutdown has occurred") may well be a Low priority alarm. This is perfectly acceptable!

Which is the more important alarm? Which would you rather respond to?

> **"Mommy, I need to gooooooo..."**
>
> or
>
> **"Mommy, I went."**

At the *pre-trip* point, the operator can still take effective action to avoid the trip, which may have still-avoidable major consequences. These consequences can no longer be avoided once the trip occurs. The consequences at that point are "how much worse" the trip-caused upset will be if the operator fails to take the correct *post-trip* actions. And as a result, the priority may well be lower, and properly so.

12.8 Safety System (ESD) Bypass/Testing Alarms

Temporary bypassing of safety system interlocks, or the signals into those interlocks, is often done for testing purposes. Rigorous control of the testing process is important. This complex subject is covered by standards that are outside the scope of this book. From the Alarm Management perspective any such bypassing must be known to the operator and must be done in accordance with site policies on such bypassing and testing.

When inputs or outputs to an ESD system are bypassed for testing, such a condition should be alarmed (Low priority) and displayed to the operator on the appropriate graphic. When testing is complete, the restoration of the interlock to its normal operating condition should clear the alarms.

New facilities must be designed and implemented to work within any relevant procedural requirements and follow the same methodology for ensuring that the operator is knowledgeable about the bypassing.

12.9 External Devices Alarms—An Engineering Love Story

"The engineer is the key figure in the material progress of the world. It is his engineering that makes a reality of the potential value of science by translating scientific knowledge into tools, resources, energy and labor to bring them into the service of man... To make contributions of this kind the engineer requires the imagination to visualize the needs of society and to appreciate what is possible as well as the technological and broad social age understanding to bring his vision to reality."

—Sir Eric Ashby

"Engineers are all basically high-functioning autistics who have no idea how normal people do stuff."

—Cory Doctorow

Besides simple sensors, often quite complex devices are interfaced to a DCS. Typical examples include analyzers, compressor anti-surge controllers, PLCs, and other similar computational devices. In such devices, many tasks are performed and much data is available for transport to the DCS. So much data, in fact, that serial or network data transfer is used rather than individual wiring for each value.

The manufacturers of these devices, of course, know them inside and out. It is their "reason for being," and in the cycle of making, installing, and improving these things over many years the devices become almost like their children, and are loved and cherished! ("Remember the Model AE-35? Boy, what a great machine. I loved the color scheme on the power supply....") Everything that occurs inside the device is, from the manufacturer's viewpoint, extremely important and needs to be available to the customer. It is common for these systems to have multiple internal health and status indications, all available for alarming.

Now, along comes the design engineer with a need for such a device. The job will require evaluating options from many suppliers, comparing long lists of specifications and characteristics, and making a final decision. All of this activity has a typical (for an engineer) result—the engineer falls in love with the selected device almost as much as the manufacturer! And when it comes time to configure the data transfer, and particularly the alarms from the device—well, they are all important! So they get all configured as alarms. Probably not well explained, but configured.

Now we have the board operator. He is NOT in love with the device; he just wants it to work, preferably quietly and without a lot of fuss. Instead, he gets cryptic and obscure alarms from it. And so, the perhaps important diagnostic alarms are buried with miscellaneous alarms about internal status ("Internal Change from Equation A to Equation B!") and none are well understood.

The proper principles to follow for configuring health and status alarms from external devices are as follows:

Configure all such alarms from the viewpoint of the operator. Remember, the alarm system is not the engineer's dumping ground for miscellaneous, nice-to-know status information! When an external device diagnostic alarm comes in, the operator needs to:

- Understand the new limitations of the connected device relative to the alarm produced
    - Is the device dead, dying, or still functional?
    - Can the readings be trusted or are they suspect?
    - Is the device just less accurate or are its readings totally unreliable?
    - Will this situation remain as it is or get worse?
- Act accordingly per Operating Procedures. Example: The analyzer is no longer functioning, so begin manual sampling
- Involve Maintenance or Staff as appropriate per procedure, based on the particular problem

The operator needs documentation about the advice related to the alarms. This does not mean an appendix in an owner's manual in the engineer's desk! Provide detailed, well-organized diagnostic displays that show the status of all of the health and status indicator inputs. Notes on the graphic can show the purpose of each indicator; don't rely on the operator to have such obscure things memorized. The graphics should also indicate the functional groups to contact for repair, based on the failure type.

Proper alarm configuration is to provide a single "common trouble" point indicating an "OR" from several status inputs. This common point is alarmed for the operator. Grouping the status points into more than one, but still a small number, of logically-related Common Trouble points is also acceptable. For example, multiple vibration instruments on a compressor should be combined into a Common Vibration Trouble Point based upon any of them reaching a particular value. All the points are shown on the diagnostic graphic, but only the Common Trouble point is alarmed. Additionally there could be an Oil System Common Trouble point being fed from several oil-related inputs.

The relevant diagnostic display should then be the associated display for the common trouble point.

The individual status points feeding the common point can usually be configured to send an event record only into the journal, if it is desired to record their individual time of activation, without their becoming another annunciated alarm to the operator.

12.10 Alarms from Fieldbus Devices

The introduction of networked field sensors and actuators, with greatly expanded internal diagnostic capabilities, poses an alarm problem. Such devices have the capability of producing many new alarm types—even preventative maintenance items such as "Time to change the oil." These are directly analogous to the prior section (External Devices Alarms) and the same guidelines for their configuration should be followed.

In new systems with many such devices, the diagnostics should be routed to and monitored directly by control system maintenance personnel, not the console operator. The operator should be notified only of diagnostics where operator action is required or where the capability of the device is compromised.

12.11 Alarm Generation by Programs

In general, a program implemented in a DCS does a task that the operator relies on. Failure of that program means the operator must act in a different way—very similar to the treatment of an external device health alarm.

Programs, however, often produce cryptic or unexplained alarms or error messages, some of which are useful to the program creator but not to the operator. Which is the better alarm?

"ERROR AT STEP 157"

or

"FLOW VALVE WILL NOT OPEN—BATCH CANNOT PROCEED"

These are real examples. Yes, you could train (and periodically retrain) the operators as to what "ERROR AT STEP 157" means, or provide a program listing and hope they will figure it out. (Author's note: They won't. Instead they will gleefully call you at 1AM on Saturday night. I know this to be true. And I deserved it.) The better solution is to have the programmer create clear alarms.

Programs must be accompanied by documentation for the operator regarding the action to take based on the specific alarm, preferably available on the graphic that shows the functioning of the program. (What, you don't have one? Then create one.)

12.12 Alarms To Initiate Manual Tasks

Some processing equipment requires periodic manual tasks to be accomplished. An alarm is often used for notification that the task must be initiated. For example, "stuff" may occasionally be routed to a unit dump tank. Over many weeks, the tank fills up. The emptying of the tank involves scheduling out a vacuum truck, often requiring 24-hour or greater notice.

An alarm at 60% of tank level might be configured to notify the operator that pumpout must be scheduled.

Even though such action is not needed in a 30-minute time frame to avoid consequences, the alarm is still needed. The priority of such alarms shall be Low. It is also possible (indeed, preferable) to put these notifications into an *Operator Alert* system (see Chapter 11, Section 6) since as alarms they are likely to become stale for the days it may take for the tank to be emptied.

12.13 DCS System Status Alarms

Alarms specific to the internal workings of a DCS system (redundant cable faults, module errors, communication errors, etc.) should be absent under normal operating conditions, and they should not be tolerated when they occur. You don't just drive around with the "Check Engine" light on! DCS status displays should have no "stale," long-term error messages. If many system status errors are present, new system alarms are difficult to recognize and respond to.

Can a single program-generated alarm cost one billion dollars?

The culmination of thousands of person-years of work and billions of dollars has arrived. The Lunar Lander *Eagle* is 40,000 feet above the lunar surface and descending rapidly. The landing profile calculations in the *Eagle's* computer disagree with the values being produced by the on-board radar. Before Buzz Aldrin can initiate the computer command to give priority to the radar, the Master Alarm annunciates—with a Code *1202* computer software alarm.

The 1202 alarm has never been encountered in any of the hundreds of Apollo 11 simulations. The astronauts do not know its precise meaning or severity, nor does anyone on the consoles in Mission Control. They all *do* know that if the computer fails during descent, they must abort the landing. Frantic queries to the back room of experts produce an answer—data rates to the computer are so high that it is overloaded and on the verge of restart. Only if the alarm remains intermittent, and not steady, can the landing proceed.

And it does remain intermittent for the next few minutes, and the Lander descends to only 3000 feet above the surface—less than a minute from landing—and then a *1201* alarm is generated! The experts are primed and ready to interpret this one—"Same type; we're Go." And history is made: the first lunar landing is made with only 20 seconds'-worth of fuel left before the abort limit.

Consider: given the thousands of things that could go wrong, and the unimaginable pressure on the astronauts, could a worse, more obscure, more distracting alarm design have *possibly* been created? It was fortunate that the specific area of the program generating these alarms had been under scrutiny due to planning simulations for the next mission. Had the meaning of these cryptic alarms taken only a few more seconds to determine, the landing would likely have been aborted.

The 3-volume set, *A Man On the Moon* by Andrew Chaiken, details this incident within a comprehensive coverage of the entire Apollo program. This set is a highly recommended addition to anyone's library.

Operators should know the proper response to each type of system status alarm, including the functional group to contact and the degree of urgency for the contact. These can be very infrequent and very cryptic when they occur, so a good list or graphic should be available to explain them.

12.14 Point and Program References to Alarms

During a D&R you will decide to eliminate many configured alarms. When it comes time to implement the decisions made in the D&R, there are some very serious problems to watch out for. There are some poor (but common) DCS programming/configuration practices that can have serious consequences if they are not dealt with correctly. These practices involve programming the DCS to take actions based specifically on alarm behavior.

For example, there may be a DCS logic interlock to make pumps start or stop based on low level in a tank.

Poor Practice: Configure the logic block with the "occurrence of the alarm" (often via a flag) as the input to cause the pump action to occur. This is poor because the alarm trip point parameter, or even the existence of the alarm, is subject to change from a variety of places. Plus, in some DCSs you have many "obscure" choices as to suppression options on an alarm, some of which could negate the flag that you chose to trigger the pump.

If you had such a logic block, and then eliminated or changed the alarms on the tank, you would alter the performance of the logic action—or even possibly eliminate it, without knowing about it.

Better practice: Configure such logic blocks with the Process Value (PV) as an input, and compare it to a numeric contained in the logic block. This is better because even though the numeric could be changed, logic blocks are more "obscure" and people are not nearly as likely to go digging around in them and changing things. The logic will activate based on the PV, whether the alarm does or not.

The better practice leaves the flexibility for reconfiguring the alarm as appropriate without inadvertently changing the performance characteristics of the interlock.

Not only DCS logic points must be checked, but also programs and signals into PLC logic and other similar things. We have found these poor programming practices to be pretty common.

Our conclusion is that, if you want something to happen based on the process attaining a certain value, then program it or configure it based on reading the value itself, not on whether an alarm occurs or not at that value.

12.15 Operator Messaging Systems

The principles covered in this book apply to some aspects of a DCS other than the alarm system, particularly any type of operator messaging system used. If the messaging system attracts the operator's attention by sounding tones or flashing lights, and requires acknowledgement, then the messaging system has a similar effect as the alarm system in loading the operator. Therefore, the use of such messages should meet many of the same principles as alarms.

Operator messaging systems were originally provided for use in batch production processes. They allow batch sequential programs to prompt the operator to do any manual steps (physical field operations, decisions, enter lab results) necessary to move the sequential operations of the batch forward. When invoked, a confirmation option required the operator to confirm that the manual steps had been completed so that the program could resume. They were also used to notify (without requiring confirmation) the operator of significant batch milestones (e.g., end of batch).

Use of a messaging system other than status annunciation or confirmation prompting is beyond the original intent and should be avoided. There are other ways to announce status to the operator; for example, graphic elements that display sequence status without generating messages. Only status changes that require operator confirmation (i.e., action, such as entering needed data) before proceeding should use messages.

An example of a common misuse of messages would be that one sequence has successfully completed and the next sequence has started, as per plan or "normal" conditions. A message or alarm should indicate abnormal, not normal conditions. The operator is better served by having a graphic that shows sequence state and progress rather than individual messages coming in reflecting normal progress. The operator will ignore the entire message system if it mostly announces that everything is progressing OK.

The usual worst case scenario if the operator message is ignored should be delayed production, not a process condition that will worsen. Alarms should be used for process conditions that will worsen if ignored. Operator messages should not. Some DCS vendors allow more than one message priority. There should be no need to assign a higher priority to a message. If a condition has more severe consequences or should be responded to more quickly, then the condition should be alarmed and not sent via the messaging system.

Any messaging system should use a separate visual and audible interface (different tones) than the alarm system.

CHAPTER 13

Common Alarm Problems and How To Solve Them

"I have yet to see any problem, however complicated, which, when you looked at it in the right way, did not become still more complicated."

—Poul Anderson

In this chapter, we will review several categories of common, problem alarms and provide several excellent techniques for dealing with them. Such alarms are often called "nuisance" or "bad actor" alarms. Bad actor alarms represent those alarms that operators categorize as a nuisance or as meaningless. With enough bad actors, an alarm system is rendered virtually useless. This may lead to hazardous plant conditions, since important or critical alarms are lost in the "sea" of bad actor alarms. This situation can have adverse economic, environmental, or safety consequences.

We usually see that the "top 20" most frequent alarms comprise anywhere from 25% to 95% of the entire system load. Obviously, if those alarms are dealt with successfully, then major system improvement will occur, and with comparatively little effort. It is quite amazing that such high numbers of bad actor alarms exist—because it is doubtful that the best control engineer in a company could actually design alarms to behave in the ways we will discuss. Yet they do exist; all varieties are in almost every system we analyze.

An additional adverse consequence of bad actors pertains to resource loading on the DCS communications network. Each annunciation, acknowledgement, and clearing of a bad actor alarm causes unnecessary network traffic and use of limited hard disk space for journaling. The reduction of available network resources is particularly distressing during unit upsets when bad actors and "real" alarms flood the network. DCS systems do not have infinite bandwidth! Or even very large bandwidth, in some cases.

The three primary tools we will use to correct bad actor alarms are:

- Proper alarm deadband configuration
- Proper process filter configuration (with warnings!)
- Proper alarm delay time (on-delay or off-delay) configuration

You may be familiar with the first two, but not the third. You will see that delay time analysis and adjustment is one of the most powerful techniques in existence for dealing with certain types of nuisance alarms.

13.1 Alarm Bad Actor Resolution

The techniques in this chapter can yield very significant results. Figure 13–1 shows some examples.

On the systems in Figure 13–1, less than 50 points each were analyzed by the techniques in this chapter. Each system shows Recorded and Annunciated alarms, because for this particular DCS there is a method (usually abused) of nuisance alarm suppression used, that sends the alarm event only into the journal, and not to the operator.

Wouldn't you be pleased if you analyzed about 30 alarms and cut your alarm rate by more than half? Here's how.

REC – Recorded Alarms ANN – Annunciated Alarms	Baseline Total Alarms:	Reduction That Would Have Occurred by Implementation of These Recommendations	Overall Percentage Reduction
System 1 – REC. ALARMS	339,521	325,423	96%
System 1 – ANN. ALARMS	291,972	280,671	95%
System 2 – REC. ALARMS	225,668	133,307	59%
System 2 – ANN. ALARMS	60,591	30,161	50%
System 3 – REC. ALARMS	414,887	333,395	80%
System 3 – ANN. ALARMS	173,721	132,924	77%
System 4 – REC. ALARMS	93,848	71,372	76%
System 4 – ANN. ALARMS	7,190	1,932	27%
System 5 – REC. ALARMS	79,434	72,935	92%
System 5 – ANN. ALARMS	4,749	1,717	36%
System 6 – REC. ALARMS	64,695	46,749	72%
System 6 – ANN. ALARMS	20,746	4,252	21%

Figure 13–1 Improvement amounts from alarm bad actor resolution.

13.2 Chattering and Fleeting Alarms

Imagine an alarm that cycles between actuating and clearing three or more times per minute. This is the definition we will use for a chattering alarm. Obviously, the clearing of the alarm condition is not due to an operator's detecting it, analyzing the situation, and making a change in the process that then moves the process value and thus results in the alarm clearing! In fact, if you asked your best control engineer to *design* you an alarm that would cycle in and out 20 or 30 times per minute, they would probably be at a loss as to how to do that! Yet, such chattering alarms are quite common. They are a big nuisance and distraction to the operator. They are relatively easy to fix.

Both analog values and digital (on-off) signals, such as from switches, can and do chatter, with digital values typically being the worst case.

There is a sub-category of alarms similar to chattering, called fleeting. These are alarms that come in and clear very quickly (too quickly for the operator to have been responsible), but do not necessarily repeat. The methods for addressing chattering alarms will also deal with these.

For analog sensors that chatter, the first thing to consider is the deadband of the alarm. Here is a brief review of the deadband concept.

13.3 Alarm Deadband

Deadband and On-Off Control

On-off control is the most basic form of control. A certain degree of deadband is placed around the setpoint. If the process variable is lower (or higher) than the deadband, the control action is turned on (or off) (see Figure 13–2).

On-off control is commonly used for regulating temperature of non-critical processes, such as central heating and air conditioning, lubrication oil temperature, cold-room temperature, etc. It can be used on a pump to fill (or empty) a tank (on at 20%, off at 80%, similar to a toilet tank).

With On-Off control, the process variable will always be cycling through the deadband. The only way to decrease the amplitude of cycling is to reduce the width of the deadband. The resulting side effect is that the frequency of the oscillations will be increased, which will reduce the life of the final control element (i.e., relays, control valves, etc.)

Early WWI Control Problems

On-off control would not work well for automobile cruise control. Imagine turning the engine on and off to regulate your speed rather than having a continuously variable throttle. But, believe it or not, some early World War I biplane engines worked exactly that way! No throttle, the engine ran wide open all the time, and there was an ignition cutout switch that the pilot would use to "blip" the engine for reduced power for landing.

Some WWI airplanes had rotary engines. This is not the Wankel-type as in some Mazda automobiles. This means that the entire engine rotated! The crankshaft was bolted to the firewall, the propeller was fixed to the engine, and the entire engine-propeller combination rotated around the fixed crankshaft. This placed a large rotating mass—like a giant gyroscope—at the front of a very light airframe with small control surfaces. Turns in one direction were very different than turns in the other, and there were other strange (by modern standards) aircraft handling characteristics. Range was limited not only by fuel but by oil—oil was pumped one-way from the engine center and then slung out of the rotating mass, not recirculated (hence the reason for the scarf—it wasn't for fashion). If you were a WWI aviator, you probably felt that your plane was trying to kill you as much as the enemy was.

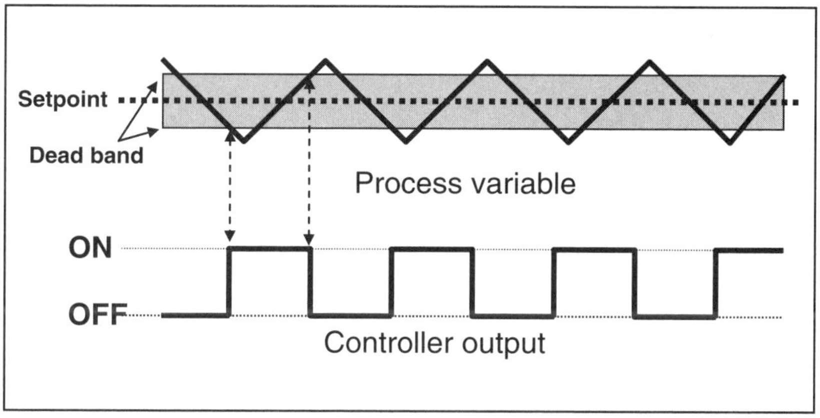

Figure 13–2 Deadband and on-off control.

Deadband and Alarms

Similar to deadband for setpoints and process control, alarms on analog values should also have a deadband specified. As a process value passes through an alarm setpoint, any "noise" or slight variation of the signal will cause multiple alarms, if there is too small of an alarm deadband.

Figure 13–3 shows how a proper deadband, larger than the noise in the signal, reduces alarm events as the process value moves above a High Alarm trip point. The alarm deadband should be specified to be larger than any expected signal noise. Most DCSs allow for deadband, but may specify it in measurement units, or percent of range, or in some other way. The deadband's plus or minus positioning relative to the trip point may also vary based on the alarm type. You have to check the DCS documentation to configure deadband properly.

Deadband should be configured on every analog alarm. Rigorous calculation is not usually necessary; the following good "starting values" can be used. (See Table 13–1.)

13.4 Process Value Filtering and Alarms

It is possible to filter process variable signals in a DCS, usually in a variety of ways. The primary reason to use filters has to do with control loop performance, not with alarms. Noisy signals act to interfere with good control loop performance. An optimum filter setting is one that smoothes out signal noise, but has little effect on the desired response of the system. If a filter is too large, it may obscure process problems from the operator.

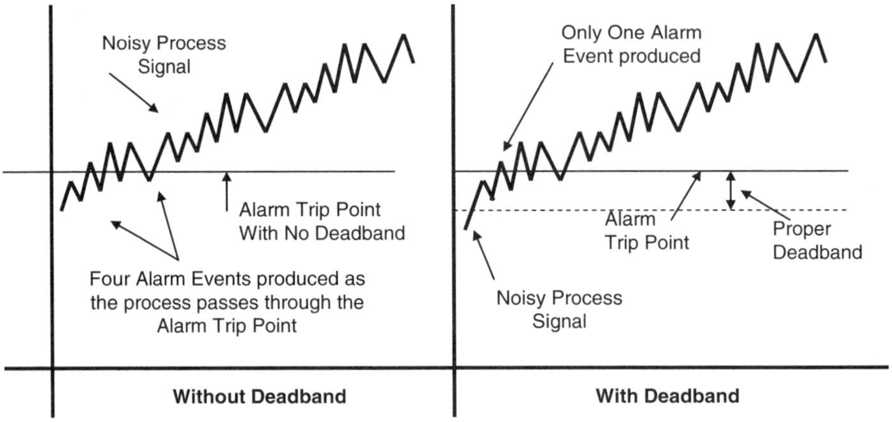

Figure 13–3 Deadband and alarms.

Table 13–1 Deadband Settings Based on Sensor Type

SIGNAL TYPE	Deadband
Flow	5%
Level	5%
Pressure	2%
Temperature	1%

Filters also introduce additional lag into control loops, which will be seen as additional apparent dead time in the loop. This may have a significant effect on the loop's settling time. A controller always has to be retuned after adding or modifying a filter on the process variable.

The determination of proper filter settings for control is generally a full chapter in a control engineering textbook. It is more important that process value filters act correctly for control than for their resulting alarm characteristics. Therefore, we do not advocate signal filtering as a good way to address chattering analog alarms.

If you suspect that you have a control problem related to a noisy signal, the values in Table 13–2 are seen as good starting points for filters.

Table 13–2 Filter Time Constants Based on Sensor Type

SIGNAL TYPE	Filter Time Constant
Flow	2 seconds
Level	2 seconds
Pressure	1 second
Temperature	-none-

13.5 Delay Time Analysis and Alarms

"Delay is preferable to error."

—Thomas Jefferson

Deadband and process filtering are applicable only to analog values. Often, the worst case chattering alarms or fleeting alarms are associated with on-off signals, such as pressure and level switches. While these devices may have a screwdriver-type, trial-and-error "deadband" adjustment on them, there is a better method to use, one that is probably already a capability of your DCS. This method applies to both analog and digital point types.

The method and technique require a bit of explaining, but once explained, the technique itself is simple. And the results you will get are so powerful that it is well worth the effort!

There are two types of alarm delays available in many DCSs, namely the ON-delay and the OFF-delay. Some alarm types may have either delay available, some only one of them. Again, there is that need to read the documentation! These are sometimes referred to as "debounce timers."

Since alarm analysis software is needed to improve an alarm system, let's make some use of it in not only detecting, but solving our problems. To do that, we want to take one of our nuisance alarms and perform two frequency analyses on it. These are analyses of the time-in-alarm (duration) and time-between-alarms (interval).

Time-in-Alarm and Time-Between-Alarms

DCSs produce time-stamped event records of at least three things. These are the alarm event itself, the "return-to-normal" event, created when the condition that caused the alarm to occur has cleared, and the operator acknowledgement event, created when the operator hits the acknowledge key for the alarm. Only the first two are of interest to us here.

You will have recorded, in your alarm analysis software, thousands of events from your nuisance alarms. For each specific nuisance alarm, take each "pair" of alarm events and return-to-normal events, and then subtract the timestamps. The result is the time-in-alarm (duration) of that alarm event. You will find that for many of your most frequent and chattering alarms, this duration is often only several seconds in length.

In a similar method, subtracting an alarm return timestamp from the next alarm event timestamp produces the time-between-alarms (interval). Plot the results for thousands of events and you will often see a graph similar to those seen in Figure 13–4.

In the case shown, based on thousands of samples, most of the alarms from this point have durations (solid line) less than 10 seconds and the time-between-alarms (dotted line) is mostly less than 20 seconds. Obviously an alarm that comes in, lasts less than 10 seconds, then goes away all by itself does not meet the basic criteria for an alarm—something that requires operator action!

This is very powerful information to use, when coupled with the ON-delay and OFF-delay abilities of the DCS. Here is exactly how those abilities work.

ON-Delay: Use of the ON-delay time parameter can prevent a transient alarm from ever being seen by the operator. The alarm must remain in effect longer than the time specified before it is initially annunciated at all. The conversation goes like this:

> *Nuisance Alarm to Control System:* "Hey! I am in alarm!"
>
> *Control System:* "Yeah, I get that a lot from you. I am going to wait—let's see—10 seconds before I tell the operator."
>
> *Nuisance Alarm to Control System, 8 seconds later:* "Hey! I am not in alarm anymore!"
>
> *Control System:* "Yeah, I figured that would happen. Good thing I didn't bother the operator with your prior message."

Only if the alarm lasts more than, in this case, 10 seconds without clearing does the Control System annunciate the alarm to the operator.

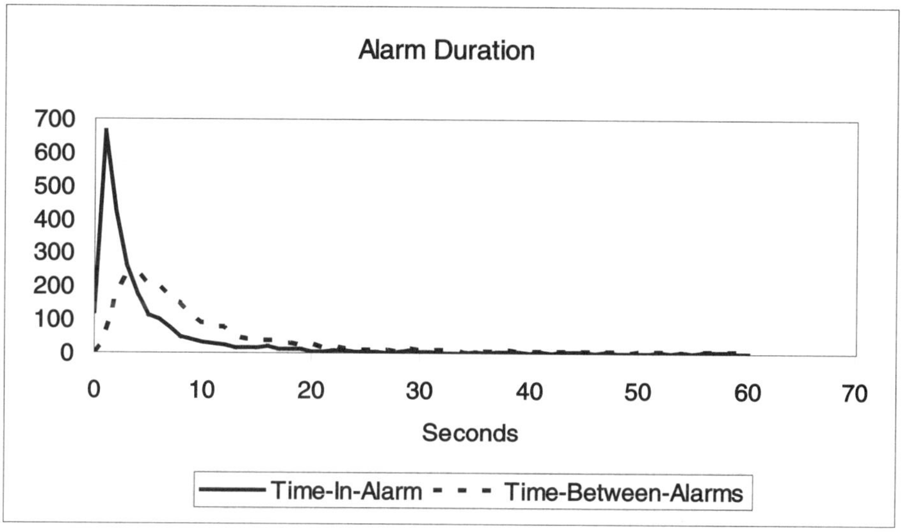

Figure 13–4 Alarm delay time analysis graph.

The correct choosing of the ON-delay time parameter is quite important since, if used, even a valid alarm is not immediately presented to the operator, which will increase the overall time it takes for a proper response to be made. This delay could be a safety concern on some points. ON-delays of 30 seconds or less are generally not a problem for Low priority and High priority alarms. ON-delays of more than 30 seconds or a minute should be applied with much care.

OFF-Delay: Use of this powerful method can turn a string of repetitive, nuisance, chattering alarms into a single, longer-duration alarm event. The OFF-delay timer must expire before a return-to-normal signal is processed and the alarm cleared. The conversation goes like this:

> *Nuisance Alarm to Control System:* "Hey! I am in alarm!"
>
> *Control System:* "Right! I am telling the operator immediately!
>
> *Nuisance Alarm to Control System, 11 seconds later:* "Hey! I am not in alarm anymore!"
>
> *Control System:* "Yeah, I figured that would happen. Look, I am not going to tell the operator that you have cleared, because I think I know what you will tell me next. I am starting the 20 second timer on you."

Nuisance Alarm to Control System, 18 seconds later: "Hey! I am in alarm!"

Control System: "Hah! I was right! Good thing I didn't bother telling the operator that you had cleared. I will leave you in alarm."

Only when the alarm has remained cleared for more than 20 seconds (in this case) does the system process that to the operator.

Using this technique, hundreds or thousands of nuisance alarm events can become a single, longer-duration alarm event. The key is the correct choosing of the delay time parameter to be greater than the normal time-between-alarms.

The disadvantage to this technique also concerns the delay time. If the operator gets the alarm and takes a corrective action that will eliminate it, he will not see a return-to-normal condition until after the delay time has expired, regardless as to if the action was immediately successful. In most cases this is quite acceptable, for OFF-delays of up to a couple of minutes. Additionally, for points that chatter for long periods, the resulting alarm will become stale, which is still a far preferable condition to chattering!

For each alarm, based on the analysis results, it is straightforward to generate a table like Table 13–3. This numerical analysis yields the exact percentage of how many alarms would have been eliminated based on the choice and type of delay time. The table (and chart) lets you see where the diminishing returns are, and to pick your delay correctly.

For this alarm, an ON-delay of 30 seconds would eliminate over 96% of the events. An OFF-delay of 1 minute would eliminate 72%.

The table used actual event data to determine the proper value. When implementing new points, the following starting values (per EEMUA) as seen in Table 13–4 may be helpful.

Both of these methods are "fixes" or "workarounds"—they address the behavior of the alarm without determining the root cause as to why the signal is chattering. Each type of input hardware (such as a switch) may have different causes and compensation mechanisms. This technique does have the benefit of immediately addressing the chattering behavior, without suppressing the view of the alarm entirely from the operator, as alarm suppression would do.

While appropriate use of the methods can dramatically improve alarm performance, the underlying process and mechanical causes should also

Table 13–3 Delay Time Alarm Reduction Table

Delay in Seconds	**% Reduction**	
	Time In Alarm (ON-Delay)	**Time Between Alarms (OFF-Delay)**
5	77.7	19.7
10	87.6	37.8
15	93.0	48.7
20	95.4	58.4
25	96.1	62.4
30	***96.5***	64.1
35	97.6	66.5
40	97.8	68.7
45	97.9	69.6
50	98.2	70.6
55	98.5	71.6
60	98.5	***72.2***
65	98.6	72.4
70	98.7	73.2
75	98.7	73.6
80	98.7	74.1
85	98.7	74.6
90	98.9	75.1
95	99.0	75.7
100	99.0	75.8
105	99.0	76.0
110	99.2	76.4
115	99.2	76.9
120	99.2	77.2

Table 13–4 Recommended Delay Times Based on Signal Type

SIGNAL TYPE	Delay Time (ON or OFF)
Flow	15 Seconds
Level	60 seconds (use >30 with care!)
Pressure	15 seconds
Temperature	60 seconds (use >30 with care!)
Other	5 seconds

be investigated—which often involves review of sensors and installations. That is, of course, if you have the time, money, or people available to do that. Many places do not, and a great Band-Aid applied is sometimes a problem solved!

13.6 Most Frequent Alarms

These are the "most frequent" alarms, regardless if the specific criteria for chattering or fleeting are met or not. As covered in the Baseline chapter, usually a very small number of alarms create most of the system load. Direct your improvement efforts to where they will do the most good via the techniques in this section.

13.7 Stale (Long-Standing) Alarms

"No guest is so welcome in a friend's house that he will not become a nuisance after three days."

—Titus Maccius Plautus

Stale alarms come in and remain in alarm for extended periods (>24 hours is a good value to use). They are often caused by stable unit conditions, such as equipment shutdown. They distract the operator by filling up the alarm summary screens. Stale alarms can only be dealt with by an understanding of the process states and hardware involved. They are usually eliminated by reconfiguring them so that they truly reflect only abnormal, unexpected conditions requiring operator action. This may require some imagination, or implementation of some state-based alarm methodologies.

13.8 Duplicate Alarms

"You see, Mr. Bond, there are two of me."

—Blofeld

Configured Duplicate Alarms

Interconnections between points in a DCS can create cases of duplicate alarm configuration. As an example, a measurement may be sent from a sensor point to a controller point, to a totalizer point, to a logic point, and so forth. Often a "bad measurement" alarm is configured on each point, and thus if the sensor points goes into that condition, several simultaneous alarms will result. These distract the operator by annunciating multiple alarms caused from a single event. There should only be one such alarm configured, on the point where the operator is most likely to take the action. If a controller is involved, this (and not the sensor point) is the proper place, since the action to be taken from a bad reading is likely to put the controller in manual and adjust the output.

Dynamic Duplicate Alarms

Duplicate alarms are alarms that persistently occur within a short time period of other alarms. If you use your alarm analysis software to list the alarms that always occur within one second of each other, you will likely find a good list to work on. Such alarms are highly likely to be multiple annunciations, in different ways, of the same process event—an undesirable situation. The individual situation will determine which are kept and which are not, or what adjustments must be made. A high quantity of potential duplicates shows the need for rationalization to eliminate them.

13.9 Nuisance "Bad Measurement" Alarms

It is quite surprising to see the amount of alarm events on most systems, of the type representing a "bad measurement" (see Figure 13–5). These are often in the hundreds or thousands!

When the loop was designed, did someone tell the control engineer the following? "Oh, and by the way, I want this sensor to go into 'Bad Measurement' under the following (several) conditions, and I want 650 'Bad Measurement' alarms per week at a minimum." And, if that had been told to the best control engineer in the company, could they have done it? Probably not! Yet, we find these on almost every system we look at.

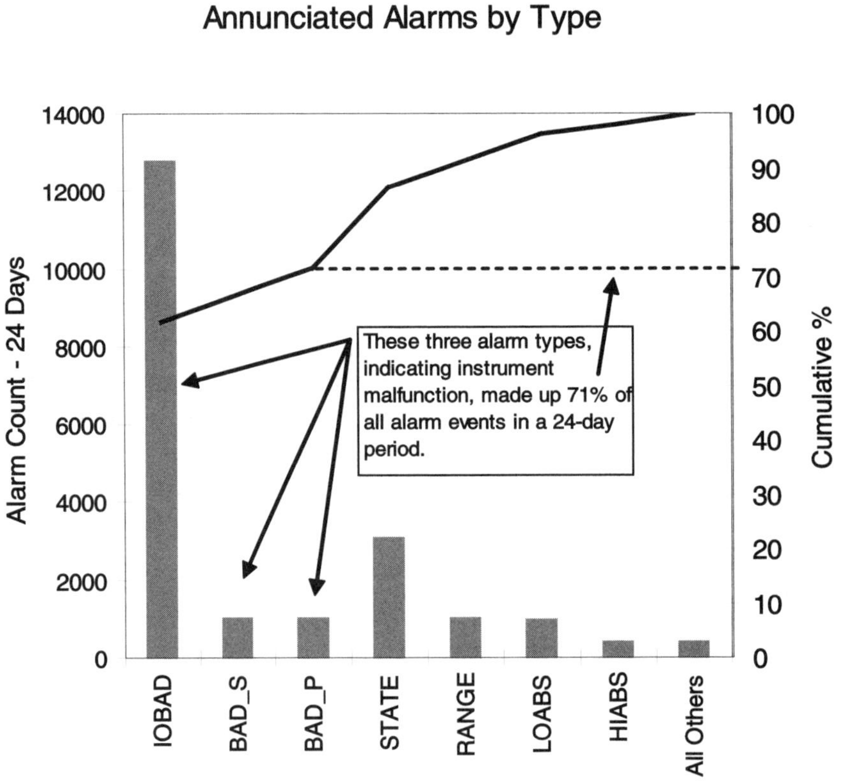

Figure 13–5 Example alarm system dominated by bad measurement.

Since no instrument was designed to be in such a state, every one of these situations can be fixed, and they should not be "just tolerated." They are mis-configured in range, or in "measurement clamping", or there is an installation problem (impulse leads filling up, etc.) The original justification for installing a flow meter probably did not include a specification that it was OK if it didn't work half of the time! If that had been proposed, the money would have never been spent to buy it in the first place.

These situations must be addressed in a prompt manner since often an instrument malfunction removes an identified, rationalized indicator of an abnormal situation from the operator's view. The time that operators spend confirming the instrument problem reduces their attention to other operator duties.

Generally the addition of a new instrument must follow a MOC methodology, to ensure it is done properly. So does the removal of an instrument, to ensure that it is truly not needed and the removal is done properly. And functionally, the indefinite toleration of a malfunctioning instrument is the same as removing it. If there is an incident, it will be difficult to explain how a relevant instrument was allowed to malfunction for months—to effectively be removed from service—without the appropriate level of review. This is the stuff of fines and lawsuits.

Long ago, the available instrument sensors had a significant trade-off between accuracy (significant digits) and range; you could obtain high accuracy only over a small range, probably less than the possible variation of the process. Control engineers were well aware of this trade-off and were accustomed to designing within those constraints. Now, along comes the digital electronic revolution and these old constraints can usually be thrown out of the window! Modern sensors can generally provide all of the accuracy needed over the entire range that the process is likely to vary. But some engineers continue to follow the older configuration practices and do not consider the consequences of generating lots of Bad Value alarms during conditions such as startup and shutdown.

The default should now be to configure the range for the entire possibility of the process and see if the accuracy you get is enough. If not (rarely, with modern transmitters), buy a better transmitter! But don't configure where you know you will get Bad Measurement conditions at ambient or shutdown conditions.

Differential pressure flows are often the worst offender. If, at zero flow, there is a slight imbalance in the leads, the meter attempts to report a slight backwards or negative flow. The flow range might not be configured for a slight negative, so the Bad Measurement condition and alarm occurs. Such points should be configured to handle that case. A cutoff can be configured and clamped at a zero value, so a small negative flow number is not actually used—which could affect some downstream calculations.

Most DCSs have the ability to "clamp" an analog value at the end of the range rather than go into a Bad Measurement state. This ability should be fully understood and used properly. (Darn! More reading of the documentation!) Controller points using the value will usually have "shed modes" —predetermined actions to take when the measurement goes bad.

13.10 Summary

Nuisance alarms can be dealt with in several ways. Dealing with a very few alarms, in the ways we have shown, can create a large and easily calculated improvement in an alarm system.

Chapter 14

Management of Change of Alarm Systems

"A computer lets you make more mistakes faster than any invention in human history, with the possible exceptions of handguns and tequila."

—Mitch Ratcliffe

Management of Change (MOC) is a well-understood and required manner of doing business in the process industries. Any addition of piping, instruments, equipment, or even tubing is rigorously controlled with documentation, safety checklists, training requirements, and approvals. This methodology, however, often overlooks the DCS alarm system. DCS changes are very easy to accomplish. Basically, you walk up to a keyboard and, with the right knowledge and access level, simply type in a change. Because of this, the changes sometimes fly "under the radar" of MOC scrutiny. If this situation exists at a facility, it must not be allowed to continue.

To maintain the integrity of the alarm system, Management of Change (MOC) procedures that address changes to alarm systems must be in effect. Such changes must be properly evaluated, authorized and communicated to all affected personnel and shifts.

In following the recommendations of this book, you will have expended much time and effort in determining and implementing proper alarm configuration. Without a proper MOC system, the configuration will drift into a suboptimal set over a short period of time. (Guaranteed!)

The following changes must be controlled with a proper MOC system. The change system itself must be designed to accommodate the number of changes that are necessary, without an over-burden of paperwork and without compromising safety.

- Changes in alarm priority
- Changes in alarm trip point
- Creation of new alarms
- Deletion of existing alarms
- Change of alarm type
- Change of alarm description or text message
- Suppression of alarms
- Point execution status (turning a sensor "on" or "off")
- Changes in alarm presentation on graphics
- Additions of, modifications to, or updates to alarm handling capabilities such as Alarm Shelving systems or State-Based Alarming configuration

While not specifically about alarms, the following changes should be controlled as well, in a way to ensure that only authorized, knowledgeable people perform the changes.

- Controller tuning parameters
- Point ranges
- Modification of logic points, interlocks, embedded programs, DCS operating system software, and similar functions

Alarm Audit and Enforcement Software should be used to periodically check for changes from the proper settings (in the Master Alarm Database), to report such changes, and to restore the system to the proper settings.

The alarm system champion for the area in question should be notified of all alarm changes. The proper alarm settings reside in a Master Alarm Database. The MOC system must ensure timely update of that database so that proper changes do not get "undone" by the enforcement software.

MOC is not required for the operation of approved, installed alarm handling strategies such as state-based alarming, flood suppression, or alarm shelving. Alterations to the configuration of these strategies themselves, however, must be done utilizing MOC with proper review and authorization.

Chapter 15

The Future of Alarm Management

"Where there is no vision, the people perish."

—Proverbs 29:18

15.1 Where is the iPOD of Process Control Industries?

On February 19, 1878, Thomas A. Edison received patent no. 200,521 for a phonograph. Edison tested his new invention by singing the nursery rhyme "Mary Had A Little Lamb" into the phonograph. The machine played back the words of Edison, and for the first time in history, man was able to record and replay sound.

About 100 years later, in 1979, Sony introduced the Walkman, a miniature portable cassette tape player, and forever changed the landscape of personal entertainment. It took about one hundred years to advance from the invention of sound recording technology to the portable stereo system. The Walkman offered about an hour of music on two sides of a cassette.

Just over 20 years later, in 2001, the genius of Steve Jobs and his team of innovators at Apple introduced the iPOD, a portable digital media player with enough storage to capture an entire collection of music and photo albums. With an iPOD, one can play hundreds of hours of music without having to listen to the same song twice. The fidelity of the stereo sound of an iPOD competes with that of the most expensive sound system and is as close to perfect as human ears can detect.

The process control industry seems to have taken a parallel path, at least in timeline, to that of the sound recording technology. The first industrial automatic controller is believed to have been invented in the 1870's. Electromechanical, pneumatic controllers and mainframe computers dominated the process control industry until 1975 when Honeywell introduced the first distributed control system (DCS). Since the advent of DCS, incremental enhancements have been made to all levels of automation.

The most significant breakthroughs yielding the most return on investments in process automation have materialized in advanced process control and digitization of business work processes through application of information technology. The resulting technologies have allowed manufacturers to increase throughput, enhance product quality, and improve personnel productivity. As a result, operating companies have been able to reduce operating costs by headcount reductions of unprecedented magnitude.

An unintended consequence of implementing best-in-class automation systems and information technology has been information overload for the operator. Lack of interoperability between layered applications and compartmentalized focus on individual layers has created a silo effect that has led to what some consider a confusion of systems and applications in the automation world.

What is needed today is significant breakthrough in automation technology that will consolidate existing best-in-class automation solutions in a seamless environment with significant emphasis on operator effectiveness.

The resulting solution can bring the *sophistication* of the most advanced solutions with the *simplicity* of an iPOD. This is the vision to aim for.

15.2 A Vision for Alarm Management

"A Vision For Alarm Management" must focus on what is yet to come and not the technologies that are available today. As with any other vision, this one stretches the imagination and assumes that current boundaries will be expanded and technological obstacles will be eliminated.

Discussing a vision for alarm management without addressing the broader issue of operator effectiveness, plant safety, reliability, and profitability is symptomatic and similar to fighting a fever in a patient without treating the body as a whole; it is incomplete.

Early in this book, we established the definition of an alarm as an event requiring operator action. Therefore, in solving the root cause of the alarm management problem we must focus on why certain events require operator intervention. The purpose of a modern control system is to run the process automatically and require only input from the operator or a supervisory (e.g., advanced process control) program. Problems arise when the control system fails to achieve the desired process behavior due to reasons such as equipment failures, uncontrollable process disturbances, or acts of God.

Eliminate the need for operator intervention and the need for alarms disappears. Automation is the foundation for eliminating the need for operator intervention. Scenarios requiring operator intervention include the following examples:

1. Move the process from one state to another
2. Equipment failure
3. Control system shortcomings
4. Errors from human actions
5. Acts of God such as invasive lightning and sudden change in temperature

The first of the five items above is a necessary element of agile manufacturing. While items 2-4 may appear to be different in characteristic, they can actually be grouped together as one—*human error,* which includes mistakes, oversights and negligence. Acts of God are a reality and must be taken into consideration when designing process technologies and process automation strategies.

Process automation strategies of the future will minimize human intervention in the same manner the commercial airline industry has leveraged automation to optimize the reliability, safety and fuel efficiency of modern airplanes. Key characteristics of future process automation strategies include the following:

- Single and seamlessly integrated software platform utilizing a common plant data model that assures data integrity

- Intelligent field instrumentation with sensor validation, control loop health monitoring, diagnostics and intelligent role-based alarm notification that engage the console operator only when absolutely necessary
- Equipment health monitoring with intelligent role-based early notification system to notify equipment maintenance personnel before problems materialize
- Early fault detection leveraging rigorous models, empirical models, and conditional logic to alert the operator of emerging process upsets
- Integrated profit optimization, transition management and reliability system
    - Dynamic model based, adaptive multivariable predictive controls integrated with transition management system.
    - Automated transition management for:
        - Startups
        - Shutdowns
        - Feed rate, product mix, and composition change, etc.
    - Automated procedures integrated with MPC and transition management systems
- Real-time knowledge management system
- Integrated intelligent operator interface with rules that will make it difficult for the operator to miss an event or make a mistake when entering data
- Intelligent alarm management that requires operator intervention well in advance of an event only when the automation system cannot independently address an emerging problem

Once implemented properly, the only time an operator will receive an alarm in this environment is when an event has occurred that cannot be solved by the automation system. Inherently, alarms will become a rare occurrence that will appear very infrequently, once or twice a week; and when they do occur, the operator will indeed take action.

While many of the individual applications required to realize the automation system described above exist today, the disparate nature of these systems, and the high cost of implementation makes the effort a monumental one.

What is needed is more detailed definition of the automation strategy of the future, investment in research, and ultimately, outside-the-box innovation and bold creativity to bring this vision to reality. We predict the process automation of the future, when realized, will be the "iPOD" of the processing industries.

Although this final chapter is labeled "A Vision for Alarm Management," we predict there will be a minimum role for alarms in the future of process automation. We believe the alarm problem of today results from the abuse of well-intentioned tools provided by existing systems. In the future, effective process automation systems must deal with alarms as an effective tool and not neglect its role in the big picture.

APPENDIX A

EFFECTIVE OPERATOR GRAPHICS

DCS operating graphic displays should act to always and effectively help the operator control the process in the best possible way. Custom graphics are the most common method for conveying process information on a DCS. The proper design of such graphics is a book all by itself. We will summarize here a few basic principles, with several "do's and don'ts."

The function of a control system graphics package is to provide a clear and understandable window to the process, with easy, intuitive, functional controls. Great care should be taken to provide the best format possible to display necessary process information and controls in a logical manner. The best way to operate involves active surveillance of the process, lots of display changing, and lots of examination of trends. Graphic design must support and promote this method. A very wrong way to operate is to wait for an alarm then figure out what is happening!

Control system graphics are used and monitored by operations, engineers, and maintenance personnel as much as 24 hours per day. Ergonomic considerations are important. Effective graphics have the following characteristics:

- Clarity—Graphics are easy to read and understand, and show the process flow clearly. They are organized to minimize information overload to operations. Unnecessary information and clutter are eliminated.
- Functionality—The operator must make decisions and control adjustments to the process, therefore ease of control manipulation is important. The functionality of the control graphics should be standardized, intuitive, straightforward, and involve minimum keystrokes or pointer manipulations. The objective is that the operator can concentrate on the process while intuitively operating the control system.
- A hierarchy of information prominence where alarms are the most prominent element. Measurement and control information (measurement values, controller modes, valve positions, etc.) are less prominent, but are more prominent than process lines, vessel

shapes, etc. The graphic will effectively draw the operator's eyes to the most important information.

- Consistency—Graphic elements and controls must behave and function consistently in all graphics and situations. Failure to accomplish consistency can significantly interfere with proper corrective action during abnormal situations.
- Navigation is logical and performance-oriented.
- Ergonomics—Operations personnel will use these graphics constantly. User fatigue must be minimized. This requires that:
  - Color conflicts are avoided. Background and foreground colors are chosen that allow for easy readability.
  - Symmetrical designs and layouts are used to minimize eye fatigue.
  - Graphic call-up speed is fast, always meeting user expectations.

A.1 Graphics Principles

Graphics Should Look Boring!

A graphic that is optimally designed for running a process and handling critical conditions effectively will, in fact, look boring. Here are some examples of what we mean.

Poor Graphics have:

- Big flashing flames that show when a burner is on.
- Brightly colored process vessels rendered with 3-D shadowing, 3D process lines and pumps.
- Spinning agitators and pumps, moving conveyors, splashing liquids and sprayers.
- Attempted color coding of process piping with their contents.
- Measurement units (psig, gpm, etc.) spelled out in big, bright text.
- Liquid levels in vessels displayed in bright colors the full width of the vessel.
- No trends.

- An exact representation of the P&ID with every minor connection and manual valve.
- Lots of lines that cross.
- Process flow from left to right, right to left, top to bottom, and bottom to top.
- Limited, haphazard navigation from screen to screen.
- Inconsistent color coding of various elements.

You get the idea. Avoid the kinds of graphics that look great and impressive when kids come to the control room to visit on Career Day, or when the corporate V.P. comes by on a tour. You might impress those folks, but the operator that uses them every day will be using a very poor set of tools to accomplish their primary job—operating the process safely and profitably.

Good graphics have:

- Gray backgrounds to minimize glare.
- No animation.
- Black process lines, with major lines shown slightly thicker.
- Limited use of color, and alarm colors are used only to display alarms and *nothing* else. If yellow is an alarm color, then yellow is never used as a text label, line color, border, or any other non-alarm-related element.
- A layout wherein the process flow is from left to right whenever possible. Gas should flow up and liquids flow down.
- A layout that is generally consistent with the operator's mental model of the process. The operators may relate better to the physical plant layout than the layout as shown on a P&ID. They may always enter the process from the south, so an equipment layout as perceived from the south might be optimum.
- Logical and consistent navigation methods.
- Display access with a maximum of three operator actions—"mouse clicks" or keystrokes.
- Techniques used to minimize the possibility of operator mistakes, and provide validation and security measures. For example, a

graphic element pushbutton to initiate an infrequent shutdown action should include a step to confirm the operator intentions. Major process upsets have occurred by mistyping an input—for example, opening a slide valve to 47% instead of 4.7%! DCSs using membrane keyboards are particularly susceptible to this type of error.

- Proper layout to avoid unnecessarily crossing lines.
- Measurement units shown in low contrast lettering, if used at all (the operators know the units of measurement.)

Use a Proper Hierarchy of Graphics

A concept of hierarchy (Levels) should be followed in constructing DCS displays. The primary purpose of these levels is to provide different levels of operating detail to aid the operator in performing different tasks. A secondary purpose of these levels is to allow easier navigation.

These levels are referred to as Level 1, Level 2, and Level 3 for the purposes of this specification.

The three levels of displays represent increasing levels of complexity. As a result of display hierarchy, a single Level 1 is associated with several Level 2 graphics that are in turn associated with several Level 3 graphics. Each Level 2 can only be associated with a single Level 1 and each Level 3 can only be associated with a single Level 2.

The majority of operator actions should be taken in Level 2 and Level 3 graphics, as shown below.

Level 1—Process Area Overview Displays

This display will show the broadest available view of the facilities under a single operator's control. It is a "big picture" at-a-glance view of the process unit. The Overview display incorporates elements such as:

- Key controllers, alarms, trends, messages
- Indication of the most significant alarms
- Key performance metrics such as Safety, Environmental, Economic, and Quality

- Key information from upstream and downstream units
- Advanced control mechanisms performance and status

Overview displays are not necessarily a pictorial. They may be mostly a tabular listing of important status and summary information, with trends to help provide the "big picture."

Level 2—Process Unit Control Displays

Level 2 graphic displays are designed to serve a dual purpose:

- They provide information regarding key elements of a specific plant unit.
- They allow operators to execute common/critical abnormal situation interventions.

The Level 2 graphic contains all the information and controls required to perform most operator tasks associated with a specific plant unit, from a single graphic.

All controllers and important indicators are shown on the various Level 2 displays. The displays are used for routine tasks such as manipulating controllers, operating pumps, starting blowers, opening valves, and so on.

Level 3—Process Unit Detail and Support Displays

All control loops (controllers, indicators, alarms, status switches, etc.) are shown on the Level 3 displays. They are also used for detailed investigations and interventions that are not time-critical. The Level 3 displays also provide support screens, such as interlock details, diagnostics, help displays, calculation details, and so forth.

These detailed displays are mainly intended for troubleshooting or manipulating items that are not accessible from the Level 2 displays.

Level 3 Support displays include:

- Custom pre-built trend displays
- ESD diagnostic displays
- Interlock Diagnostics and trouble shooting displays
- Common Alarm displays with details of sensor status

- Detailed information about equipment, instrumentation
- Detailed status of Advanced Process Control functionality
- System—supplied displays such as Point Detail, System Diagnostics, etc.

Other displays at Level 3 not associated with the display of live process information can include the integration of:

- Operating Procedures
- Alarm Documentation and Response Guidance
- Abnormal Situation response guidance
- APC documentation and operational procedures
- Other program-related documentation and operational procedures
- And similar items of use to the operator

Use LOTS and LOTS of Trends

Every Level 2 and 3 graphic should contain *at least* one trend of the most important process values depicted. The best operators actually run from trends; they don't need pictures of the distillation column showing each tray.

In fact, for each process area, you should create Level 2 custom displays consisting of nothing but the trends and the controllers of the most important values to watch. The trends should have all of the controls needed to alter the time base (to show the last 10 minutes, or 2 hours, or the last 24 hours, for example). When the graphic is called up, the trend's Y-Axis span should automatically range itself to a predetermined scale appropriate to the process conditions, or even relative to the current value of the reading. This is *rarely* the full scale of the value! It should be a tight scale where meaningful change of the value is immediately detectable. The operator should not have to manipulate a lot of keys to make the trend usable.

A proper trend of a controller will show Process Value, setpoint, and controller output. Such trends are highly useful for diagnosing controller problems and, particularly, valve problems. Studies have shown that, on average:

- More than half of all control valves are incorrectly designed, with a nonlinear installed valve characteristic and/or incorrect sizing for the application
- More than 50% of all control valves have mechanical problems of Hysteresis or Stiction (see sidebar at the end of this chapter)
- Almost 20% of control strategies can be improved upon
- 10% to 20% of industrial control loops are run in manual
- About one third of controllers produce more variability in auto than in manual!

Now you may say, "But this DCS has the ability to pick any point and trend it if the operator wants to." And you would be right, they can. But they won't, not if it takes 10 or 20 keystrokes and mouse clicks to generate them. Operators have different skill levels with the DCS functions, and it's your job to see that the right tools are on the system so that *every* operator can run the process effectively.

Navigation and Functionality

The DCS operator interface system shall be designed to minimize the number of keystrokes required to identify, verify, and assess an alarm.

Every point with a configured alarm should have an associated graphic display on the DCS. This associated display should aid the operator in the proper diagnosis and mitigation of the event that caused the alarm. Methods by which the operator is quickly directed with a single keystroke or button-click (i.e., one-touch access) to the associated display should be used. Most DCSs have this capability, but it must be configured.

Multiple methods of navigation should be provided. The operator should be able to go up and down through the hierarchy, and side to side through the process, and call related details, trends, and shutdown status displays from any graphic. This navigation capability should work with all available methods provided for navigation by the DCS vendor—mouse or touch screen target selections, keyboard keystrokes, context sensitive menus, etc.

The system and graphics should be configured such that it is *never* necessary for the operator to type in a point name or graphic name. The ability to get to any graphic without knowing the hierarchy should be available for maintenance and engineering users of the system—this may include an overall menu or direct entry of graphic names that are displayed/printed with the graphic.

A single alarm interface should be used, namely that of the DCS. If alarms can come from sources nominally "outside" of the DCS, those should be brought into the DCS, if the DCS is used in any way to respond to the alarm. All alarms should be acknowledged only once; it should never be required to acknowledge the same alarm in more than one place.

Graphics should not be "hard-coded" with alarm behavior for points; the behavior should be consistent based on the configuration of a point's alarm and should change if the configuration changes. Example: If the practice is that an alarm is shown in a different color based on priority, the graphic should detect the priority currently in effect on the alarm. It is a common practice to "blink" or "flash" unacknowledged alarms, and then have them go "steady" when acknowledged but still in effect. Do this with care; the process value indication itself should not blink, since that puts the item you are interested in out of view! Instead, the background of the value or a box around it can blink.

A process graphic must visually and consistently identify tags in alarm, whether or not the alarm is acknowledged, and the priority of the alarm.

A.2 Hardwired Switches

For some critical installations, console mounted switches may be provided to allow the board operator immediate access, or to meet a required degree of redundancy. These must be carefully designed to enhance the overall abnormal situation management by the board operator. In all cases, they must be placed in a logical spatial pattern, in a consistent manner, be well-labeled, and be integrated to allow a degree of history collection when actuated.

Hardwired switches are often used so that there is no issue with the operator being able to activate them at a moment's notice. Physical switches can be emulated via software constructs in the DCS (such as graphic button actions). Emergency and/or bypass switches placed via software in the DCS should utilize graphic techniques to ensure activation by an operator within a specified time frame as required by operations.

Implementation of switches in software is often preferable, being less expensive and making console layout much simpler. When this is done, it is important to use the proper graphic and programming techniques so that the software switch will be easy to access and activate—and not be hidden by other windows or "buried' many screen-levels deep in a modal graphic mechanism. The implementation of software techniques for switches must take into account that there is generally an expected speed-of-actuation for such a switch.

For hardwired/soft switches activated only by the board operator, no alarms should be generated directly by the activation of the switch. For hardwired switches that may be actuated at multiple places (which may include the board), it may be appropriate to alarm the activation of non-board switches to ensure that the board operator knows of such actuation and makes the proper response.

A.3 Hysteresis and Stiction—The Most Common Control Valve Problems

While slightly off of the Alarm Management topic, these phenomena should be known and understood by anyone involved in operating a process. So, a short explanation.

Hysteresis

Imagine that you are Dr. Frankenstein. A great lightning storm tonight is forecast, and you have high hopes that, *this time*, you will succeed with Frank Jr. Unfortunately you have to do all the work yourself tonight, because Igor is drunk again. (It is hard to get good help around here.) So you go to the control panel and push up the big lever that raises the platform, containing a moribund Frank Jr., up from the basement to the laboratory level. When the platform rises to the right height for working on him, where you can conveniently hook up those electrodes, you mark the position of the lever with a pencil.

What's that? Thunder? Yes! You push the lever up to its top position, and the platform rises out the top of the laboratory, exposing Frank Jr. to the full fury of the electrical storm! Which, unfortunately, never gets closer than 10 miles. Another night wasted!

Darn, now you have to disconnect the electrodes. You pull the lever down to your pencil mark. But the platform doesn't come all the way down to where it was. It stops 2 feet higher! The platform control has hysteresis! The platform (valve) actually moves to a different position given the exact same control lever position (controller output), depending on the *direction* of the platform (valve) movement—up or down.

Well, you'll fix that platform control later. Probably need to adjust the tuning coefficients. *(Author's note: The Doctor is mistaken, you cannot correct for hysteresis by adjusting controller tuning coefficients, but you would be amazed at how many operators and engineers that think they can! A loop containing a valve with hysteresis or stiction cannot be properly tuned for optimum control response. The mechanical problem must be addressed.)*

Stiction

Now you need to go to the loading dock and bring in those extra "parts" that were delivered this afternoon. Ah, they were left in a big cardboard box on the concrete floor, and all you have to do is push them into the freezer room.

So you get behind the box and push. Nothing. You push harder. Nothing. You push really hard and suddenly, the box slides three feet all at once and you fall down on your bad knee. Ouch! You think about creative ways to punish Igor if he ever sobers up enough to notice.

The cardboard box on the concrete floor had *static friction* or *stiction.* It took much more force to get it moving than was needed to keep it moving, and the box overshot the desired response. On a control valve, this shows up when you increase the output signal, and "nothing happens." You increase it more and still, nothing happens. You increase it still more and finally something happens—and generally, way too much happens.

Proper trends of process value, controller setpoint, and controller output can show these conditions of hysteresis and stiction.

Appendix B

An Example Alarm Philosophy

How do you take all of the best practice information provided so far and put it in a coherent, concise Alarm Philosophy document? This appendix provides guidance. The book is intended to provide lots of background and explanative discussion about each topic, but your alarm philosophy should not be so lengthy. However, it is also important to not condense your philosophy too far, as many engineers will be prone to do!

Remember that a major purpose of the philosophy document is that it be used as guidance for people that are unfamiliar with the topic of alarm management. These people may be in project roles directly involved with alarm implementations. Enough detail must be included to give such people the education they need so that their alarm implementations will be proper and consistent. This guidance strikes the correct balance.

"Best Practice" Alarm Philosophy Table of Contents

1.0 Alarm Philosophy Introduction
2.0 Purpose and Use of an Alarm Philosophy
3.0 Alarm Definition and Criteria
4.0 Alarm Annunciation and Response
- 4.1 Navigation and Alarm Response
- 4.2 Use of External Annunciators
- 4.3 Hardwired Switches
- 4.4 Annunciated Alarm Priority

5.0 Alarm System Performance
- 5.1 Alarm System Champion
- 5.2 Alarm System Key Performance Indicators (KPIs)
- 5.3 Alarm Performance Report

6.0 Alarm Handling Methods
- 6.1 Nuisance Alarms
- 6.2 Alarm Shelving
- 6.3 State-Based or State-Dependent Alarms
- 6.4 Alarm Flood Suppression
- 6.5 Operator Alert Systems

1.0 Alarm Philosophy Introduction (Example)

This document serves as a guideline for the development, implementation, and modification of alarms at DCS board operator positions at <<<name and location of facility>>>. These guidelines should provide an optimum basis for alarm selection, priority setting, and configuration, in order to promote safety while minimizing duplication, noise, and confusion.

2.0 Purpose and Use of an Alarm Philosophy (Example)

The goal of this Alarm Management Philosophy Document is to ensure that the DCS alarm system acts as a tool to always and effectively help the operator take the correct action at the correct time. This is true if:

- Alarms are properly chosen and implemented.
- Alarms are relevant, clear, and easy to understand.
- Alarms are configured consistently in accordance with industry best practice guidelines.
- Alarms are presented at a rate that the operator can effectively handle.
- Operators can rapidly assess the location and relative importance of all process alarms.
- Operators can process alarm information during high frequency alarm actuation events.
- Alarm systems are properly controlled, monitored, and maintained.

The Philosophy document is an alarm design guideline for both new systems and modifications to existing systems. It is for both in-house use and contractor use during projects.

The Philosophy document provides a consistent and optimum basis for

- Alarm Selection
- Priority Determination
- Alarm Configuration
- Alarm Handling Methods
- Alarm System Performance Monitoring
- Nuisance Alarm Resolution
- Alarm Detection, Presentation, and Annunciation
- Operator Interface for Alarms
- Operator Response to Alarms
- Alarm System Management of Change

The Alarm Philosophy is based on several key assumptions:

- No amount of alarm management will replace the constant surveillance of a qualified operator.
- Operators are trained on the alarm management strategy.

- Alarm Management enhances the operator's ability to make a judgment based on experience and skill.
- Operators will respond to all alarms, regardless of priority. *(Corollary: the system design therefore must not produce more alarms than the operator can respond to.)*
- Alarm priorities define the order of the operator's response.
- The alarm system is routinely maintained and kept up to date.
- Alarm management includes all categories of alarms coming to an operator, including system alarms.

3.0 Alarm Definition and Criteria

The decision to configure an alarm must meet certain criteria. The most import of these criteria is:

Does the event require operator action?

A process alarm is a mechanism for informing an operator of an abnormal process condition for which an operator action is required.

In the remainder of this section, the contents of the chapter "What Should Be an Alarm," should be included almost in its entirety, including good and poor practice examples. This is valuable information to establish the basic principles of alarm management and indicate the fundamental and necessary thought process for considering the creation of an alarm.

4.0 Alarm Annunciation and Response

Provide a detailed overview of the steps the operator must take to respond to an alarm, such as:

- Detection
- Identification
- Verification
- Acknowledgement
- Assessment
- Corrective action
- Monitoring

4.1 Navigation and Alarm Response

Provide the principles for an effective operator interface to the alarm system, as covered in the chapter, "DCS Alarm Display Capabilities," and in "Appendix A: Effective Operator Graphics." Note any DCS-specific settings that are implemented around the alarm display behavior.

4.2 Use of External Annunciators

Based upon your site policy for the use of external lightbox annunciators, incorporate the proper principles from the chapter "The Death of the Lightbox."

4.3 Hardwired Switches

Include the guidance criteria on console switches from "Appendix A: Effective Operator Graphics."

4.4 Annunciated Alarm Priority

Incorporate the guidance on the proper role of alarm priority from the chapter, "DCS Alarm Display Capabilities." Note your site choices for all priority aspects such as priority names, colors, and sounds.

5.0 Alarm System Performance

Summarize the KPIs to be monitored from the chapters "Baseline and Benchmarking of Alarm System Performance," and "Monitoring and Reporting Alarm System Performance."

Note the goal level (Overloaded, Reactive, Stable, Robust, or Predictive) to be sought for each system.

5.1 Alarm System Champion

Enumerate the duties of the Alarm System Champion, from the chapter "Monitoring and Reporting Alarm System Performance."

Specifically name the site or unit job role to be assigned those duties.

5.2 Alarm System Key Performance Indicators (KPIs)

Determine and document the goals and KPIs your systems will be targeted to achieve. Use the chapters "Baseline and Benchmarking of Alarm System Performance," and "Monitoring and Reporting Alarm System Performance."

5.3 Alarm Performance Report

Indicate the content, frequency, and distribution lists of periodic alarm system performance reports.

6.0 Alarm Handling Methods

In this section, summarize all of the issues covered in the chapter, "Real-Time Alarm Handling Methods."

6.1 Nuisance Alarms

Here, address the dangers and proper practices to be followed regarding temporary alarm suppression.

6.2 Alarm Shelving

Document the proper site methodology for Alarm Shelving.

6.3 State-Based or State-Dependent Alarms

Even if your site does not currently employ state-based alarms, cover the concept and guidelines around the practice in this section. State-based alarming is probably in your future.

6.4 Alarm Flood Suppression

Again, even if your site does not currently employ alarm flood suppression techniques, cover the concept and guidelines around the practice in this section. Alarm flood suppression is probably in your future.

6.5 Operator Alert Systems

Cover the proper principles for use of operator alert systems.

7.0 Alarm Documentation and Rationalization (D&R)

Alarm Documentation and Rationalization (D&R) is a sound, consistent, and logical methodology by which alarms are determined and prioritized. Alarms resulting from the methodology are said to be "rationalized."

This section cover all aspects of alarm rationalization in detail, based on the chapters "Alarm Documentation and Rationalization (D&R)." This should be one of the longest and most complete sections in your Alarm Philosophy Document.

Develop and include your own site rationalization grids in the following sections:

7.1 Areas of Impact and Severity of Consequences

7.2 Maximum Time for Response and Correction

7.3 Severity of Consequences and Time to Respond Matrix

Cover, in detail, the use of these grids and the specific practices around performing D&R.

7.4 Alarm Documentation

Decide on the aspects of alarm documentation that will be captured in a D&R, and note the methods by which that information will be made available to the operators.

7.5 Alarm Trip Point Selection

Document the guidance around proper alarm trip point selection during D&R.

8.0 Specific Alarm Design Considerations

Up-front decisions around alarm configuration will reduce time spent in Rationalization. Decisions can usually be made in advance around several topics:

- Sensor Malfunction or similar Bad Value alarms (existence, priority, placement)
- Pre-Alarms
- Duplicate Alarms
- External Device Health and Status Alarms (ESD, Surge Control, Analyzer, etc.)
- And others

In this section, predetermine, to the maximum extent possible, all of the alarm practices to be followed for all of the alarm types covered in the chapter "Specific Alarm Design Considerations."

8.1 Handling of Alarms from Instrument Malfunctions

8.2 Alarms for Redundant Sensors and Voting Systems

8.3 External Device Health and Status Alarms

8.4 ESD Systems

8.5 ESD Bypasses

8.6 Duplicate Alarms

8.7 Consequential Alarms

8.8 Pre-Alarms

8.9 Flammable and Toxic Gas Detectors

8.10 Safety Shower and Eyebath Actuation Alarms

8.11 Building-Related Alarms

8.12 Alarm Handling for Programs

8.13 Alarms to Initiate Manual Tasks

8.14 DCS System Status Alarms

8.15 Point and Program References to Alarms

8.16 Operator Messaging System

9.0 Management of Change

Cover in detail site requirements around the proper MOC practices to be followed for changes in the alarm system, based on the chapter "Management of Change of Alarm Systems."

10.0 Training

Implementation of an alarm philosophy, in the form of either a new installation or revision of an existing alarm configuration, requires operator and staff (engineers that deal with the alarm system) training. The training must cover several areas and include specific points.

Cover the training requirements for employees regarding the alarms system, and particularly when the alarm system is changed. See the training section in the chapter, "Alarm Documentation and Rationalization (D&R)."

11.0 Alarm Maintenance Workflow Process

Include an overview of your site policies and work processes for maintaining an alarm system. The following is a generic example.

The ability to reach maximum potential for handling alarms can only be achieved by continually monitoring, tuning, and enhancing the existing alarm strategies. A workflow process must be in place to allow the alarm system champion to continuously improve the alarm system. The figure illustrates the interdependence of the various components of the alarm management maintenance workflow process, and follows Six Sigma methodology:

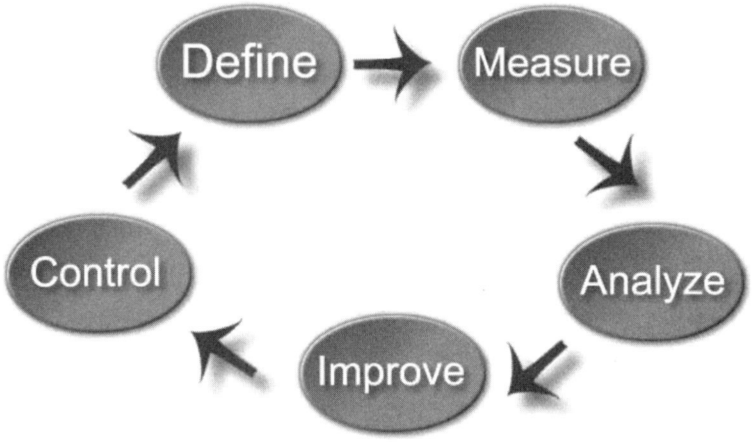

Figure B–1 Alarm management maintenance workflow process.

DEFINE
Define the Problem, Objective, and Goals of the Alarm Management effort. Decide on the desired performance level (Overloaded, Reactive, Stable, Robust, Predictive).

Develop the Project Plan and Assign Resources.

MEASURE
Develop the Data Collection and Sampling Plan, in general the proper use of Alarm Analysis software and the proper determination of Key Performance Indicators.

ANALYZE
Perform the analyses outlined in the section on Key Performance Indicators. A statistical analysis of alarm events must be used to identify trends, patterns, and bad actor alarms on the system. Initial Baseline analyses of a system are highly valuable. Publish periodic reports. Upon identifying the specific problems, perform root cause analyses, develop alternative solutions, and decide upon the proper solution.

IMPROVE
The objective of alarm analysis is to identify opportunities to improve the alarm management system. In many environments a significant number of problems can be eliminated through minor changes to the system. Based on the problem areas identified, determine and implement solutions. Any changes to the alarm management system must be carefully evaluated and planned before being implemented. Essential to successful implementation of changes is to identify and communicate the changes to the operators and other affected individuals, usually in the form of content-specific training and the standard MOC process.

The following cycles of analysis indicate whether the changes made were effective.

CONTROL
Ensure that gains are not lost over time. Properly transition the needed responsibilities from an Alarm Management Improvement "Project" organization, to the ongoing organization that will continue to use the system long-term. At a minimum, this involves continuing analysis of the alarm system performance KPIs and action plans for new deficiencies that occur.

References

Alford, J., J. Kindervater,, & R. Stankovich,. "Alarm Management for Regulated Industries." *Chemical Engineering Progress* (April 2005): 25.

American Institute of Chemical Engineers/AIChE. *Guidelines for Safe Automation of Chemical Processes*. Hoboken: Wiley Publishing, 1993.

Andow, P. "Alarm Performance Improvement During Abnormal Situations," in *HAZARDS XV: The Process, Its Safety, and the Environment: Getting it Right* (Manchester: Institute of Chemical Engineers, 2000).

Blaesi, J. "No Alarm—No Manipulation Operation Improvement for Plant Stability." Presentation, Texas A&M Instrumentation Symposium, College Station, Texas, 2004.

Bransby, M. & J. Jenkinson. *HSE Contract Research Report 166: The Management of Alarm Systems*. London: Health & Safety Executive, 1998.

Brown, D. & M. O'Donnell. "Too Much Of A Good Thing?—Alarm Management Experience In BP Oil." Presentation, IEE Colloquium on "Stemming the Alarm Flood," London, UK, June 1997.

Brown, D. "Alarm Management: A Problem Worth Taking Seriously," *Control* (July and August 1999).

Brown, D. "Horses For Courses—A Vision For Alarm Management." Presentation, IBC seminar on "Alarm Systems," London, UK, 2002.

Brown, D. "Alarm System Performance—One Size Fits All?." *Measurement+Control* (May 2003): 120.

Bullemer, P. & I. Nimmo. "Tackle Abnormal Situation Management with Better Training." *Chemical Engineering Progress* (January 1998): 43.

Cochran, E. & P. Bullemer. "Abnormal Situation Management: Not By New Technology Alone..." Presentation, AICHE 1996 Safety Conference.

Emigholz, K. "Improving the Operator's Capabilities; Observations from the Control House." Presentation, AIChE Loss Symposium, July 1995.

Engineering Equipment and Materials Users Association. *Alarm Systems: A Guide to Design, Management and Procurement.* London: The Engineering Equipment and Materials Users Association, 1999.

Errington, J. "After the Alarm Rationalization—Managing the DCS Alarm System." Presentation, 2004 Spring National Meeting, 16th Annual Ethylene Producers' Conference, New Orleans LA, Apr 25-29, 2004.

Errington, J. & P. Bullemer. "Designing for Abnormal Situation Management." Presentation, AIChE 1998 Ethylene Producers Conference, New Orleans, LA., 1998.

Errington, J. & D.V. Reising *ASM® Consortium Guidelines: Effective Alarm Management Practices.* Minneapolis: Honeywell Laboratories, 2004.

Grosdidier, P., P. Connor, B. Hollifield, S. Kulkarni. "A Path Forward for DCS Alarm Management." *Hydrocarbon Processing* (November 2003).

Health & Safety Executive. *The explosion and fires at the Texaco Refinery, Milford Haven, 24 July 94.* London: Health & Safety Executive, 1997.

Health & Safety Executive. *Better Alarm Handling* [Brochure]. London: Health & Safety Executive, 2000.

Koene, J. and H. Vedam. "Alarm Management and Rationalization." Presentation, Third International Conference on Loss Prevention, 2000.

Li, R., J. Liu, J.R. Zhou, S.L. Ling, K.W. Lim, W.K. Ho, R. Srinivasan and K.C. Tan. "An Algorithm for Reducing Repeating Nuisance Alarms in a Refinery Plant." In: The 3rd Asian Control Conference, Shanghai, 2000.

Li, R., J. Liu, J.R. Zhou, S.L. Ling, K.W. Lim, W.K. Ho, R. Srinivasan, & K.C. Tan. "Development of Practical Solutions for Alarm Management in a Refinery Plant." Presentation, The 3rd Asian Control Conference, Shanghai, 2000.

Liu, J., R. Li, S.L. Ling, J.R. Zhou, K.W. Lim, R. Srinivasan, K.C. Tan & W.K. Ho. "An Intelligent Alarm Management System in a Refinery Plant." Presentation, Chemical and Process Engineering Conference 2000 Singapore in conjunction with Regional Symposium on Chemical Engineering 2000, Singapore, 2000.

Liu, J., R. Li, S. L. Ling, J. R. Zhou, K. W. Lim, R. Srinivasan, K. C. Tan, W. K. Ho, & A. Tay. "Intelligent Alarm Management Through Suppressing Nuisance Alarms And Providing Operator Advice." Presentation, The 3rd Asian Control Conference, Shanghai, 2000.

Lyche N.P., "Alarm Management and System Design." Presentation, 7th Annual Honeywell Australasian IAC Users Group Conference, 1995.

Mattiasson, C., "The Alarm System from the Operator's Perspective," Presentation, IEE People in Control Meeting, Bath, UK, 1999.

McCulloch, J.G. "Alarm Handling and Future DCS Developments," Presentation, the Honeywell 1996 European Users Group, 1996.

Metzger, D. & R. Crowe. "Technology Enables New Alarm Management Approaches." Presentation, ISA Technical Conference, Houston, TX, 2001.

Moncur, M. & L. Moncur. The Quotations Page. *www.thequotationspage.com.* Mostia, B. "How to Perform Alarm Rationalization." *Control* (August 2003).

Nimmo, I. "Abnormal Situations Management: Giving your Control System the Ability to Cope." *Honeywell Journal for Industrial Automation & Control* (August 1995).

Nimmo, I. "Abnormal Situation Management—Adequately Address Abnormal Operations." *Chemical Engineering Progress* (September 1995).

Nimmo, I. "The Importance of Alarm Management Improvement Project." Presentation, ISA INTERKAMMA, Germany, 1999.

Nimmo, I. "It's Time to Consider Human Factors in Alarm Management." *Chemical Engineering Progress* (November 2002).

Nimmo , I. "Abnormal Situation Awareness—The Need for Good Situation Awareness." Presentation, Advances In Process Control 7, September 2004.

Nimmo, I. "Rescue Your Plant from Alarm Overload." *Chemical Processing Magazine* (January 2005).

Normen Arbeitsgen Mess Und Regeltechnik (NAMUR). *Alarm Management, Publication NA102*. Germany: NAMUR, 2003.

Occupational Safety and Health Administration. "Occupational Safety and Health Standards." Ch. 17 Section 1910.119 in *Process Safety Management of Highly Hazardous Chemicals.*

Nochur, A., H. Vedam, & J. Koene. *Alarm Performance Metrics.* Honeywell Singapore Laboratory, 2001.

Reising, D.V., & T. Montgomery. "Achieving Effective Alarm System Performance: Results of ASM® Consortium Benchmarking against the EEMUA Guide for Alarm Systems." Presentation, 20th Annual CCPS International Conference, Atlanta, GA, April 2005.

Toney, J., D. Oubre, D. Penn, A. Cheung, J. Cade. "Alarm Configuration Management: from Pilot to Product." Presentation, Honeywell Industrial Control Users Group Symposium, Phoenix, Arizona, June, 2000.

Wilson, M. "Alarm Management and Its Importance in Ensuring Safety." Presentation, IEE Colloquium on Best practices in Alarm Management, London, 1998

Index

L

M

N

O

P

R

S